Reproduction of Tactual Textures

Springer Series on Touch and Haptic Systems

Michaël Wiertlewski

Reproduction of Tactual Textures

Transducers, Mechanics and Signal
Encoding

 Springer

Michaël Wiertlewski
Department of Mechanical Engineering
Northwestern University
Evanston, IL, USA

ISSN 2192-2977   ISSN 2192-2985 (electronic)
Springer Series on Touch and Haptic Systems
ISBN 978-1-4471-6242-1   ISBN 978-1-4471-4841-8 (eBook)
DOI 10.1007/978-1-4471-4841-8
Springer London Heidelberg New York Dordrecht

A mon grand-père
A Bénédicte
A ma famille

Series Editors' Foreword

This is the sixth volume of the "Springer Series on Touch and Haptic Systems", which is published in collaboration between **Springer** and the **EuroHaptics Society**.

Reproduction of Tactual Textures is focused on the characterization and reproduction of human tactile interaction. Its growing importance is demonstrated by its application in new smart and intelligent devices, which are continuously replacing physical control keys and parts, by software components that implement the same or even more advanced functionalities. In this type of device, the tactile stimulus represents the difference between a simple graphical interface and an interactive device that properly feeds back to the user his/her action.

The majority of contributions in *Reproduction of Tactual Textures* are related to the analysis of the haptic fingertip interaction. The studies reported are based on various advanced set-ups that have been used to perform experiments about tactile perception under different conditions. Mechanical properties have been recorded in order to systematically record and reproduce vibrations, textures and tangential forces. The studies are based on the physiology of human tactile perception, which takes into account the bandwidth and displacement of mechanoreceptors that participate during the tactile interaction.

Reproduction of Tactual Textures is based on the PhD thesis of Michaël Wiertlewski who received the 2012 EuroHaptics Society PhD award. His thesis was selected as the best of many other excellent theses in the field of haptic research. In addition to the empirical studies of human tactile interaction, Dr Wiertlewski's monograph includes an excellent overview of the state of the art in tactile perception research and technologies and also represents an important contribution in this field.

Manuel Ferre
Marc Ernst
Alan Wing

Acknowledgements

I wish to express my gratitude to my friends, colleagues and mentors, who inspired this work.

In particular, I sincerely wish to thank Vincent Hayward who guided me through this work and shared his most audacious ideas. I'm grateful to José Lozada, his guidance and support have been determinant in the success of this work. My gratitude also goes to Edwige Pissaloux who believed in me from the beginning, and to Moustapha Hafez who motivated this project.

I would like to thank all the members of the Sensory and Ambient Interfaces Laboratory at CEA List. In particular, Christian Bolzmacher, Florent Souvestre, Mehdi Boukallel, Hanna Youssef, Margarita Anastasova and Christine Mégard for advice and help. I spent memorable moments with all the PhD candidates from CEA List, Hussein Sleiman, Cécile Pacoret, Emir Vela and Guillaume Trannoy. I also thank the 'nouvelle vague', Lucie Brunet, Edouard Leroy and especially Charles Hudin who collaborated on the last chapter. I thank Juan La Rivera and Corentin Roche who applied the results of this research into a working prototype. My gratitude also goes to Annick Latare and Anne Goué for helping me through the administration, and to Marc Itchah for his advised tips for designing and machining the devices presented in this book.

I'm also indebted to Yon Visell and Jonathan Platkiewicz who took the time to read and correct this manuscript. I thank all the members of the haptic group of the Institute for Robotics and Intelligent Systems, Ildar Farkhatdinov, Amir Berrezag, Abdenbi Mohand-Ousaid, Irène Fasiello, Chao Wang Chen and Rafal Pijewski, as well as Lionel Rigoux, Jérémie Naudé and Camille Dianoux for their warm welcome during the last part of the journey.

Last but not least, I thank my family and my friends for their support and my wife, Bénédicte, who was my finest ally in this adventure.

Michaël Wiertlewski

Contents

Chapter 1
Introduction

Abstract This introduction defines the scope and the content of the book. It presents the motivations of the work, and the issues related to recording and reproducing natural textures directly at the fingertip.

1.1 Introduction

If the world is at hand's reach, the hand owes it to the sense of touch. Touch allows us to become aware of our surroundings in a way that vision and audition cannot. The information collected through tactual perception is employed to identify, grasp, evaluate and manipulate objects, from tools to fruits. Touch is also a key aspect of social interactions, communication of emotions and non-verbal information transmission.

The sense of touch can be categorized into two types of sensations. One type relates to the cutaneous sense which results of the deformation of the skin on the surface of our body. The second type relates to proprioception, which relays the position of our limbs and the forces applied by our muscles to the central nervous system. The combination of both sensations is called haptic perception, from haptics, the science that studies touch.

Haptic perception grants access to a variety of physical attributes about the surfaces that are in contact with the body. These attributes comprise temperature, shape, texture, and many other attributes of the surface in contact. The transduction of these quantities into relevant messages for the central nervous system is a subject of research since the past century. In fact, unlike to vision and audition which are related to electromagnetic and air pressure waves, the physical quantities that mediate tactile perception are not well understood yet.

This book focuses on the perception of fine texture and in particular how the roughness, the small irregularities caused by the finish of the surface, is appreciated through the haptic channel. As for visual or audio stimuli, virtual tactual sensations feel more realistic when roughness is added to the raw geometry. Fine roughness, as opposite to coarse roughness, is defined through the presence of asperities that never exceed 100 μm. This scale of surface irregularities can only be experienced through the lateral motion of the finger onto the surface.

M. Wiertlewski, *Reproduction of Tactual Textures*,
Springer Series on Touch and Haptic Systems, DOI 10.1007/978-1-4471-4841-8_1,
© Springer-Verlag London 2013

The perception of texture has importance in many domains. It can be the difference between a common fabric and silk. In the case of surgery, it also gives clues as to the quality of tissues and their health. The present work aims to clarify the nature of roughness stimuli. It is motivated by sensorimotor rehabilitation, especially of post-stroke patients. After a stroke, patients often lose motor control and tactile sensibility. The rehabilitation process helps them to recover this deficit by continuously stimulating the haptic sense, especially texture sensation. But the success of recovery is limited by the fact that current exercises are not engaging and do not sufficiently motivate the patient. One outcome of this work is a device that robustly and efficiently creates a variety of virtual tactual texture and can be employed for the development of games for rehabilitation.

In addition, smartphone and touchscreen input devices have raised interest in providing feedback the user by producing tactile stimuli as a response to taps on the screen. Physical buttons are disappearing from the human-computer interaction and there is a need of communicating touch sensations and enhancing the user experience by tactile feedback. Some devices use tactile actuators to transmit high-frequency vibrations to the enclosure or the screen. Driving these actuators with rich signals can give the illusion of buzzing, impacting or even texture. The limit is that such devices cannot stimulate the kinesthetic perception.

1.2 Scope

This work aims to render realistic sensations with a single high-fidelity transducer attached to the finger. A focus has been placed on roughness simulation and rendering. In particular it explores the range of vibrotactile textures that can be reproduced through cutaneous stimulation alone. To achieve this goal, a custom made sensor records and stores the interaction of a finger with a surface. Reproduction is then made by reversing this measurement by rapidly deforming the pulp as the finger moves with the tactile transducer. Tactile exploration of fine surfaces does not rely on distributed stress on the fingertip, and even a single tactile element can render the sensations that one feels when stroking an object.

The book deals with several issues related to this goal. First, the mechatronic design of transducers dedicated to tactual texture reproduction is discussed, and a specific transducer capable of recording and rendering texture is described. Second, a mechanical measurement of the dynamic properties of the fingertip is presented. The data provide details about of the underlying mechanical phenomena that occur during tactile exploration. Lastly, an analysis of the vibrations generated by the stroke of the fingertip on several surfaces is presented.

1.3 Overview

This book is segmented in 8 chapters including this introduction and is organized as following.

Chapter 2 reviews the current literature about human touch that falls into the scope of this book. The perception and psychophysics of touch are the first aspect of this literature review. The second part is focused on the mechanical structures and properties of the fingertip and their influence on perception. The last part reviews the technological state-of-the-art for devices reproducing tactual textures.

Chapter 3 introduces a transducer designed to record and replay the interaction of a finger sliding on a rough surface. During sensing operation, tangential forces arising from the friction of a fingertip on the surface are measured with high-dynamic range and over a frequency bandwidth that encompasses tactile perception. The force-position profile is recorded and is reproduced using the same transducer in actuator mode. In this configuration, the actuator is fixed to the finger and imposes a lateral deformation of the fingertip as the finger undergoes net motion. A simple psychophysics experiment ties the roughness estimate of a real surface with the one reproduced by the apparatus.

Chapter 4 extends the previous chapter by describing in-depth the design of the apparatus in both sensing and reproduction mode. A formal model and a mechanical calibration is proposed to present the limits of the apparatus and their influence on the quality of the stimuli that are produced. This chapter also raises the question of the spatio-temporal definition of tactual roughness. In other words, it asks if the perception of texture can possibly be done by a fusion of the temporal determinant and the position of the fingertip in space. Both data are used to create spatial maps of five textures. Two psychophysics experiments on the identification of complex textures and on discrimination of gratings, try to shape an answer to this question.

Chapter 5 and Chap. 6 investigate the mechanical behavior of the fingertip to lateral deformation. This study explores two fundamental questions. The first relates to Chap. 3 where the interaction force measured is converted into displacement for the reproduction of tactual sensation. As in audio, when there is a change of units and quantity, the conversion is made accordingly to the impedance of the medium. But contrary to audio, where the impedance of the air is well defined, tactile interaction and measurement depends on the mechanical properties of the fingertip. These properties are depended of several parameters including the frequency of stimulation. The second question concerns implications of the biomechanics for tactile perception, and especially whether the biomechanics properties affect signal generations. Chapter 5 describes the instrument that was developed for the studies. It proposes a novel way of measuring mechanical impedance by combining electromechanical measurement techniques and a feedback control loop for enhancing the sensitivity of the device. Chapter 6 presents the results of a set of measurements on seven participants. Results reveal a dual mode behavior with an elastic response at low frequencies (<100 Hz) and a viscous behavior at higher frequencies. The resulting data are also valuable for informing the design of better transducers and reproduction algorithms.

Chapter 7 explores tactile interaction from a signal processing point of view. The interaction of a fingertip with several sinusoidal and flat surfaces was recorded with the same apparatus as in the Chaps. 3 and 4. Despite the large variations from one measurement to another, some invariants are extracted. A spatial representation and Fourier analysis reveals that every texture induced a $1/f$ noise. Moreover, touching sinusoidal textures produces complex waveforms that follow possess a harmonic behavior. Questions are raised on the perceptual implication of such a background noise on the tactual estimation of relative speed.

Finally, Chap. 8 concludes this book with a summary of the main contributions, perspectives that arise from this work and unanswered questions.

Chapter 2
Literature Review

Abstract This chapter reviews the literature related to tactile interaction for perception of slippage and texture. The first section presents the research in psychophysics that relate to tactile perception of texture using a bare finger or through tools and devices. The next section describes the properties of the skin from a mechanical viewpoint. Both mechanical response and friction properties will be related to roughness perception and the creation of vibrations. Finally, the current state of the art in devices that reproduce virtual haptic textures is portrayed.

2.1 Human Perception of Tactual Texture

Tactual texture concerns the surface and material properties that our finger perceives by coming into contact with an object and/or sliding on the surface. The nature of surface contact is a key factor in the tactile identification of objects. It reveals mechanical properties such as friction, roughness and temperature, and informs the central nervous system about the qualities of the contact. Furthermore, this information is essential to grasping and manipulating our environment.

2.1.1 Texture Sensing

2.1.1.1 Active Touch for Perception

Motion is an intrinsic part of haptic perception. Without making contact, manipulating, or stroking an object, it is impossible to sense the shapes, textures, materials that surround us. Katz, in his classic monograph [71, 78], was the first to notice that tactile perception is intimately linked with self-motion. He hypothesized that the size of the perceptual window is augmented by moving it in the world. The motion of the finger or the hand makes the resolution and the reach of touch virtually infinite. The same observation has been also applied to visual perception, in which the movement of the eye plays a important part in scene comprehension. Katz also claimed that without relative lateral motion between an object and the skin, the roughness of

M. Wiertlewski, *Reproduction of Tactual Textures*,
Springer Series on Touch and Haptic Systems, DOI 10.1007/978-1-4471-4841-8_2,
© Springer-Verlag London 2013

a surface cannot be estimated. Roughness is defined by micro-scale asperities of the surface; at length scale well below the spatial detection threshold of touch.

These observations have been experimentally verified by Meenes and Zigler [101] who showed that the roughness of various grades of paper is optimally perceived by moving relative to the surface. The lateral motion introduces temporal variation in the mechanical pressure applied to the skin, which is used as a cue for roughness estimation. Lederman and Klatzky [84], in a seminal article, drew a taxonomy of hand movements used for tactual perception. They found eight exploratory procedures from which humans extract information about a tangible object through of touch. Texture, for instance, is perceived through lateral motion between the object and the finger, and compliance with a variable pressure normal to the surface.

Gibson observed that vision and touch share similarities, as self-motion is necessary for perception [43]. Using simple planar shapes, he asked two groups to match their tactile sensations with drawings of the shapes. In the first group, the shapes were simply pressed into the hands of participants. The second group was free to explore the shapes and boundaries of the objects before giving an answer. The res\-ults of this experiment showed a clear advantage of exploratory motion for the perception of shapes.

As Katz noted, tactual abilities for texture discrimination are also greatly impaired by the absence of relative lateral motion. Psychophysical evaluations of human ability to discriminate coarse gratings (groove size greater than 100 µm) reveal that the discrimination thresholds drop significantly when participants are not allowed to move their finger. During this experiment, Morley et al. [108] also noticed that exploratory procedures follow an almost sinusoidal displacement of the finger from right to left. This observation can possibly be induced by the uni\-directionnality of the rectangular grating.

In light of these findings, it is clear that in order to reproduce virtual surfaces, users have to be able to explore the virtual world with their hands. In particular, texture is sensed during lateral motion of the finger onto the surface.

2.1.1.2 Perceptual Dimensions

Texture is a term that encompasses many features of the surface being explored. Hollins et al. [52, 53] tried to reduce these features to a minimum number of descriptors. They performed a study in which they presented to non-expert participants a total of 17 tactile stimuli, such as sandpaper, wood, velvet, cork, and asked them to describe the sensations they felt with a set of adjectives. From these data they ran a multidimensional scaling (MDS) analysis and found that the perceptual space could be represented as a four-dimension Cartesian space. Major axes correspond to the description of roughness (smooth-rough) and compliance (hard-soft) of the material. The two minor axes are most likely to be frictional properties (sticky-slippery) and temperature (warm-cold). It is interesting to notice that the axes are not orthogonal to each other, indicating a cross-correlation between attributes.

Using a free-sorting procedure on 24 car seat fabrics, Picard et al. [131] found, also by mean of an MDS analysis, that the perceptual space could be divided in to three to four dimensions. However the major dimensions were soft/harsh and thin/thick, as opposed to previous studies. This difference might be explained by the fact that the participant were French speaker and that the words "soft" and "harsh" are more commonly referred to fabrics than "rough" and "smooth". This result emphasis the great influence of culture and context on the cognitive classification of sensations.

The perceptual space suggested by Hollins et al. [53] was constructed around subjective attributes which depend of the context and are not always reliable. The limited number of samples must also be taken into account. In fact, 17 samples with 4 dimensions do not create enough redundancy for correct statistics. With this limitation in mind, Bergmann-Tiest and Kappers [11] performed a free sorting study with 124 texture samples. Each of the samples was classified from its mechanical properties, such as compliance (inverse of stiffness) and roughness computed from a weighted average spectrum of the height profile. The MDS procedure, they used embedded the results in a four dimensional space. The dimensions correlated with the physical measurements but they were not completely aligned, which suggests that each perceptual dimension is based on several physical properties. Furthermore, the physical attributes of roughness and compliance described a horseshoe shape in perceptual space, which contradicts the hypothesis of a Euclidean perceptual space.

When analyzing individual dimensions, Smith and Scott [141] reported that the friction coefficients (quantified as the tangential forces divided by the normal forces) was correlated with stickiness judgments on smooth surfaces. In the case of non-smooth surfaces, roughness perception also seemed to be correlated with increasing net friction (i.e., tangential force) [143]. From this viewpoint, it is clear that the continuum rough/smooth is not orthogonal with sticky/slippery. Moreover, they also found that roughness estimates were not directly correlated with topographical measurements of surface asperities. Stevens and Harris [150] performed a quantitative study of the perceived roughness of textiles with increasing emery cloth grid number; the grid number is inversely proportional to the particle diameter. They found that roughness was a positive power of grid number. They also found that smoothness varied similarly, but with opposite power, with grid number. This last result confirms that smoothness is the reciprocal of roughness. Ekman et al. [35] extended these results and found that the exponent of the power-law depends on the coefficient of friction and the material used (sandpaper, cardboard and paper).

Yoshioka et al. [185] explored the perceptual dimensions of texture explored through a rigid probe. They asked participants to rate various textures on a three dimensional space that comprised roughness, hardness and stickiness. They found no significant difference between the probe and the bare finger. The authors claimed that in the case of indirect touch, the three dimensions correspond to the vibration power transmitted by the tool, the compliance of the surface and the friction force acting on the probe during manual exploration. These results do not agree with the other studies, probably because the experimental procedure imposed the dimensions of the perceptual space a priori.

Despite discrepancy in the definition of tactual texture, all these studies corroborate the fact that texture is associated with multiple attributes in which roughness plays a fundamental role.

2.1.1.3 Differences Between Coarse and Fine Roughness Perception

Katz [71, 78] was also the first to distinguish between the perception of coarse and fine roughness. The former can be felt simply by pushing on the surface, whereas the small asperity size of fine surfaces require a lateral motion of the skin with respect to the surface in order to be perceived. Further experiments confirmed this hypothesis, by showing that without relative motion, surface defects below 100 μm in size can not be detected [51, 86, 157]. It is worth noting that these findings also showed that speed does not influence the perceptual estimation of roughness, while normal force increases roughness magnitude estimates. These observations show the independence of the roughness estimate with respect to temporal factors, and lead to imagine that roughness is represented by the central nervous system, in space rather than in time.

When a tactile signal is repeated for a long time, the human sensory system adapts and apparent amplitude decreases: when the signal is stopped, the sensitivity progressively recovers its function. Lederman et al. [87] asked participants to rate the roughness of coarse textured surfaces before and after inhibiting vibrotactile sensitivity with high and low frequency masking stimuli. No change in the perception of coarse textures was observed suggesting that the perception of coarse textures is not mediated by vibration. Later, Hollins et al. [51] performed the same experiment, using fine textures and found a significant decrease in discrimination abilities following adaptation to high-frequency stimuli. The results of both studies can interpreted by supporting a *duplex theory* of roughness perception; large asperities of roughness are perceived as spatial determinants, whereas fine textures (<200 μm of spatial frequency) are mainly perceived through temporal determinants, consisting in vibrations created by sliding friction between the textured surface and the finger. Moreover, vibrations are both necessary and sufficient for the perception of fine textures [55].

From the literature on texture perception, we can conclude that tactual perception of roughness is based on a combination of two perceptual cues: the spatial and temporal determinants. The first category relates to the deformation pattern of the skin in contact with a surface, and is dominant when surface defects are coarse. Temporal determinants are produced by the time-varying global deformation of the skin, which is excited during lateral motion between the skin and texture, and they prevail when texture is fine (i.e., <100 μm of height). Both of these cues are involved in the tactual perception of surface topography and friction properties. Depending on the mechanical events, the central nervous system will rely on one or the other cue.

2.1.1.4 Perception Through a Rigid Link Compared to Bare Finger Exploration

The central nervous system has the capability of extending the function of the body via the tools we hold. For instance when we are using a pen to write, the pen is integrated in our sensory system, and our mind controls the tip of the pencil rather than managing the forces and torques on the body of the tool. This is also true for tactual perception: a stick stroked on a rough surface transmits vibrations that one captures by the mechanoreceptors in the hand. But from a physiological point of view, the mechanical deformation field that mechanoreceptors embedded in the skin sense is only the pressure of the skin on the surface of the pen. So how does the central nervous system interpret the vibratory signal from the stick as being comparable the bare finger exploration?

Klatzky and Lederman [75] compared roughness judgments when subjects explored textured surfaces with a rigid probe or a rigid sheath mounted on their fingertip with judgements made when exploring with the bare finger or through a compliant glove. In the first condition, the rigid link imposes a coding of tactual perception based on vibration alone, whereas the bare finger condition gives access to both vibratory and spatial cues. The authors report that roughness estimates were greater with a rigid link. Discrimination performance was best with the bare finger, but the rigid sheath only reduced discrimination by few percent. This experiment shows the importance of vibratory cues during texture exploration. One explanation for the greater subjective magnitude when using a rigid link is that the finger and the compliant glove serve to damp the vibration content. On the other hand, the rigid link amplifies vibrations produced through collisions with edges of the surface.

In [88], the same authors investigated the influence of speed and mode of touch during tactile exploration with a probe. In the passive mode, participant's arm was attached to a table while a robot stroked sandpapers on the tip of the probe they were holding. In the active mode, subjects explored sandpaper by actively stroking the probe. The resulting data showed that speed has a larger effect during passive exploration. The authors argued that because kinesthetic information from the arm motion was absent, subjects relied on cutaneous cues to assess both speed and roughness. However, a conflict is involved, because estimating the speed of motion without kinesthetic feedback relies mostly on texture information. Changing the texture reference changes the distribution of frequency content, and biases hence biased the speed or the roughness estimate. In another publication, they also explored the effect of a rigid sheath superimposed on the fingertip on tactile perception [85]. Orientation and compliance discrimination were greatly reduced by the absence of distributed spatial information. Roughness estimates and vibrotactile sensitivity were, however, less affected by spatial masking. This supports the difference between spatial and temporal determinants and their importance to tactile perception in everyday life.

The work of Yoshioka et al. [185], presented in Sect. 2.1.1.2, also showed that the subjective scaling of texture stimuli on the roughness, hardness, and stickiness continua in probe condition are similar but not identical to the bare finger exploration mode. The cognitive classification of texture is analogous despite the difference in the mechanical signals sensed by the central nervous system in the two modes of exploration. The conclusion of this study was that the cognitive mechanisms implicated in texture perception differ for probe versus bare finger exploration. In the case of exploration with a probe, texture is reconstructed depending on the tool that participants are holding.

2.1.2 Texture and Slip Discrimination

Human capabilities of surface feature discrimination (periodic or non-periodic) have been intensively studied using psychophysical methods. Psychophysics is a branch of experimental psychology that relates perception and physical stimuli. In practice this leads to the study of verbal responses given by participants when they are stimulated in a controlled manner. Knowledge about the inputs and outputs of the human sensory system makes it possible to extrapolate its behavior. In practice, an emphasis is placed on measuring the lowest stimulus that triggers a reaction (absolute threshold) or the lowest difference of stimuli that triggers a response (difference threshold[1]) of human perception.

2.1.2.1 Texture

Johansson and LaMotte [62] determined the detection threshold of the height of a single defect in a perfectly smooth silicone wafer. They found that an edge as high as 0.85 μm could produce a sensation. Raised dot detection thresholds were determined to be 1.09, 2.94 and 5.97 μm for dots of respective diameters 602, 231 and 40 μm. Detection thresholds depend on the area of stimulation and, by extension, the number of receptors involved. Several studies have used raised dot matrices [80] or linear gratings [69, 108, 114] to study the effect of repetitive texture discrimination. For instance, Miyaoka et al. [106] used aluminum-oxide abrasive paper of varying grit, corresponding to average particle sizes of 40 to 1 μm. They determined that the smallest perceptible difference (difference threshold) in grit was about 2.4 μm.

The main criticism of these studies is that they were conducted with non-smooth topographical function, yielding gratings or surfaces that contain sharp edges and corners that can induce strong shocks and call to question the influence of height in the detection threshold. Louw et al. [91] considered the detection problem using a set of Gaussian bump profiles with a wide range of width ($0.15 < \sigma < 240$ mm). Participants were able to freely explore samples with their bare fingers. The minimum detectable height of the bump was about 1 μm, on average for the smallest

---

[1] The difference threshold relative to the stimulus intensity is called Weber fraction in honor of one of the founders of psychophysics: Ernst Heinrich Weber (1795–1878).

bump width. This value is about 10 times smaller than the size of skin cells themselves, raising questions about the physical limits of haptic perception. When the
results were plotted on a log-log scale, they were linearly aligned, such that the amplitude threshold depended on the width σ raised to the power 1.3. The maximum
value of the first derivative of height (maximum slope) is the dominant perceptual
cue for shape and texture perception.

Building on these observations, Nefs et al. [115] used sinusoidal gratings to explore perceivers ability to discriminate amplitude and spatial frequency with periodic and mathematically defined surfaces. They found the Weber fraction for amplitude discrimination to be about 10 to 15 % which yields a difference threshold
as low as 2 µm. The difference threshold for spatial period discrimination is about
11.8, 6.3 and 6.4 % for gratings of spatial period of 10, 5 and 2.5 mm respectively.
Spatial discrimination thresholds were higher than values found in [108], probably
because the latter authors used a continuous function for the grating shape. Conversely, the study of Morley et al. used gratings with sharp edges, so that collisions
between the finger and texture corners might have created mechanical events that
were easily sensed by the human sensory system.

The extremely high tactile sensitivity to surface defects is, as mentioned earlier,
surprising. In quasi-static settings, the tactile sensory system can barely discriminate two dots less than 0.5 mm apart [69], but the dynamic interaction results in
a hundred-fold increase in spatial sensitivity. It seems that this capability is not attributed to quasi-static sensory capabilities, but instead to the integration of mechanical signals felt during motion. In fact, the mechanical discrimination threshold in
quasi-static conditions[2] and the roughness estimate during active touch are not correlated [90]. It confirms the idea that the central nervous system uses special strategies to perceive fine textures. Thanks to this effort, even with relatively scattered
mechanoreceptors, the central nervous system is able to access sub-micrometric geometries.

Looking at the mechanical forces generated during active touch also gives some
insight into the interaction between the surface and the finger. Smith et al. [142]
measured tangential and normal forces (referred in this manuscript as F_t and F_n
respectively) that arise from the friction between a finger and a non-smooth surface.
Surprisingly, they found no effect of normal force on subjective scaling of the roughness. However the derivative of the tangential force (dF_t/dt) is highly correlated
with roughness estimates. Moreover the net friction force (average of the tangential
force) is also a factor in perception. Authors hypothesized that the central nervous
system assesses the roughness by comparing the root mean square of variation of
the tangential force with the net friction coefficient. Therefore a lower friction force,
achieved by lubricating for instance, results in a lower roughness estimate with the
same surface pattern.

---

[2]The spatial discrimination threshold is also known as the two-points discrimination threshold.

2.1.2.2 Slip and Velocity Perception

The awareness of slipping and the estimation of relative speed are crucial for day-to-day manual interaction. Interacting with our environment involves grasping and manipulating objects. Cutaneous information about the mechanical properties of objects helps to accomplish these tasks correctly [19, 176]. The reader can find more details about the role of cutaneous perception in precision grasping tasks in the literature [179].

Texture and slip sensations are intimately linked in human perception. To demonstrate this fact, Srinivasan et al. [147] used a perfectly smooth glass substrate stroked under a static finger. In the absence of defect on the surface, participants were not able to detect steady slip. However, a single 4 μm-high asperity on the surface was enough for participants to detect relative motion. Discrimination of the magnitude of relative velocity between the moving object and the finger is poor. Essick et al. [36] found that brushing the arm at different velocity yields to the Weber fraction is about 25 % and the scaling between actual velocity v_a and perceived one v_p is well described by $v_p \propto v_a^{0.6}$. The last results have been determined using a stimulus that had a limited duration, and therefore the actual speed could have been determined by either speed or duration. Depeault et al. [32] experimented with textured drum rotating under participants' skin and found a relation closer to $v_p \propto v_a^{1.1}$. They explained the discrepancy with the results of Essick et al. by the presence of a fixed duration of stimuli and participant could not rely on time to assess the velocity. They also explored the effect of various textures on speed judgement, and it appeared that speed estimates are varying with spatial period. One explanation is that speed is temporally encoded relative to a spatial texture, when the finger is static. Therefore, if the speed v doubles and at the same time the spatial period λ is also doubled, the temporal frequency f remains the same ($f = v/\lambda$). In this experiment the finger was stationary, so that the velocity estimate was based only on tactile information. In active perception the somatosensory system integrates the motion of the limb to make an estimate of speed. Moreover, the texture was fine, so the central nervous system can be assumed to exclusively base its estimate temporal frequency content. It is interesting to note that the reciprocal is not true; roughness estimate was independent of scanning speed [102].

Contrary to steady slip, the initial transition between stick and slip is a mechanical event perceptible even in absence of asperities on the surface [147]. This rapid transition creates vibrations that are felt by the sensory system and exploited by the central nervous system to evaluate the slipperiness and the roughness of the contacting surface [64]. This initial slip provides sufficient information for the central nervous system to be able to assess the friction coefficients of the material and to adjust the normal force accordingly [141]. Moreover the direction of slippage seems to be given by the direction of the tangential force applied on the finger during sliding [107].

In conclusion, relative velocity and slippage are key pieces of information for object identification, but are also crucial for grasp and manipulation. The central nervous system develops a number of strategies to evaluate these cues, and in both cases texture plays a determinant role. Sliding motion on a surface gives rise to mechanical vibrations that excite the human sensory system and that are used to assess contact conditions. If slippage is imminent, the body shape and muscle tonus are adjusted to avoid dropping the object. Evidence also shows that information about slip and texture is contained in variation of the tangential force over time. The net tangential force vector during incipient slip or sliding is also a primary indicator of slipperiness. The norm of the tangential force is correlated with stickiness and the friction properties of surface being touched. The direction of this vector also gives information useful for grasp stability, for instance, and sliding direction.

2.1.3 Mechanotransduction

The fovea of the retina and the fingertip share much in common, despite their numerous differences [187]. Both are two-dimensional tissues that comprise a dense population of receptors creating zones of high acuity [69]. Fingertips, which are covered by glabrous skin, embed mechanoreceptors near the surface of the skin. These cells have the property to transduce mechanical disturbances, such as pressure and vibration, into action potentials that are transmitted to the central nervous system through the nerves. Different type of mechanoreceptors populate hairy and glabrous skin, although some types are present in both. The present review is limited to receptors that are sensitive to mechanical stimulation and that are present in the hand and the glabrous skin. The reader can found additional information about the mechanisms of sensory transduction in recent literature [31].

Most of these receptors are sub-millimeter in scale and their behavior is inferred from anatomical observations and electrophysiology. The former have revealed the existence of four types of mechanoreceptors in the glabrous skin. All of them are connected to myelinated nerve endings that transmit the electrical impulses to the brain. The latter has enabled researchers to probe ulnar and median nerves that run in the arms, and to directly record the action potentials thanks to a probe clamped on the outer of the nerve fiber. The capture of electrical signals during mechanical stimulation of the skin reveals four types of afferent messages. They are classified according to two attributes: the speed of adaptation to mechanical events (*slowly adaptive* SA and *rapidly adaptive* RA) and the size and the sharpness of their receptive field (type I for small receptive field with sharp edges and II for large and blurry bounded field). By habit, RA I afferents are named simply RA and type II are called PC from their well identified connections with Pacinian corpuscles.

2.1.3.1 Merkel Complex and SA Type I Afferents

Merkel nerve endings are SA I nerve afferents branching into multiple neurites that terminates in the vicinity of Merkel disks located on the boundary between the dermis and hypodermis. This boundary is undulated and the Merkel disks are located at the tip of each undulation. They are distributed widely in the glabrous skin, and the afferent density can reach 100 per cm^2 at the fingertip [67]. Each nerve afferent branches into up to 90 fibers with two possible architectures. Some of the branches are clustered around one region and others are distributed along long chain that can reach 200 μm of length. There is still a debate about the role of the Merkel disks and their contribution to mechanotransduction [49].

SA I afferents exhibit a discharge rate that is linear with the indentation depth of a square punch, as observed in experiments with primates [14], and explained by a power law of exponent 0.7 in humans [76]. They are known to be sensitive to spatial features, and are especially responsible for the perception of edges [130], curvature [47] and coarse texture [25, 68, 134]. When scanning coarse gratings with the skin, a population of SA I afferents responds to the spatial frequency of the grating independently of speed, suggesting a spatial encoding of the stimulus [46]. Gabor filters convolved with the spatial pattern of the surface are able to predict the neural code of the SA I afferents [25]. Even though the code is mainly spatial, dynamic touch increases the sensitivity of SA I afferents by a factor of ten. Moreover, the firing pattern is not affected by velocity [70]. SA I nerves respond to a sinusoidal motion of the skin increasing in frequency by a decrease of the number of impulses by second, which makes them insensitive to vibration [92].

The spatial sensitivity of SA I afferents and their poor temporal resolution evidence the dual theory of texture perception. Coarse textures are most likely represented by SA I afferents which respond to the edges and curvatures of the surface. Fine texture perception, however, is probably to mediated by rapidly adaptive afferents, which are more sensitive to vibrations.

2.1.3.2 Ruffini Endings and SA Type II Afferents

The role of Ruffini receptors is obscure. It was previously associated with SA II afferents [23], but their contribution has been recently put into question [121]. Only one Ruffini corpuscle has been found in the skin of the index finger using immunofluorescence despite, the fact that SA II afferents account for 15 % of all afferents in the median nerve [63]. However, SA II afferents are clearly associated with lateral stretch of the skin, and they respond to the lateral force during the incipient slip period. They also contribute to the proprioception of the hand (with Golgi tendon organs and muscle spindle receptors) by responding to the strain of the skin on the finger, thus giving an indication of finger position.

Some Ruffini-like structure connected to SA II afferents have been found at the frontier between the skin and the nail. These last afferents are correlated with large scale tangential force direction and amplitude on the fingertip [13]. They could be responsible for the sensation of stickiness during texture exploration as it has been shown that human sensory system relies on tangential forces cues for assessing the friction [141].

2.1.3.3 Meissner Corpuscles and RA Afferents

Meissner corpuscles are found in the glabrous skin, located at the interface on the dermis and the epidermis, as in the case of Merkel cells. But, unlike the latter, they are found in the grooves of ridges, and therefore are closer to the surface of the skin. They are composed of stacks of Schwann cells entangled with unmyelinated axon afferents. Conjunctive tissues link the stack to the boundary of the epidermal ridges that is supposed to act as a lever and amplify deformation of the ridges. For a detailed description see [120].

These anatomical properties (proximity to the surface and lever-like arrangement) are probably the explanation for the ultra-low threshold of RA afferents to transient mechanical deformation of the skin. In fact, a single 2 μm high dot moving under the skin is enough to stretch the papillary ridges and activate action potentials [82, 147].

These mechanoreceptors are found on each side of the fingerprint ridges. In glabrous skin the density can reach 1.5 afferents per mm^2. They respond to stimuli within a radius of 3 to 5 mm, which corresponds to a receptive field five time larger than SA I afferents. However, their sensitivity is about four times higher than for SA I afferents [67]. These properties, combined with the fact that they are insensitive to static stimulation, make them ideal candidates for the perception of slip and micro slip that can occur when lifting an object [64]. They are believed to be the source of the reflex that regulates grip force to avoid the slipping of a grasped object. See [179] for a review of tactile contribution to the control of grasp.

RA afferents respond to vibrotactile stimulation between 8 and 64 Hz [65, 110]. At high amplitude, their response is similar to that of the PC which could explain humans' high sensitivity to vibration. Nevertheless, their role in fine texture perception is not clear. Adaptation to a 10 Hz vibrotactile signal, targeted to lower the sensitivity of both SA I and RA afferents, does not reduce fine texture estimate [54]. This result implies that they are not determinant in the perception of fine texture but more specialized for transient mechanical event like slippage.

2.1.3.4 Pacinian Corpuscles and PC Afferents

Pacinian corpuscles are found more deeply in the dermis. They are ovoid-shaped cells composed of about thirty concentric lamellae separated by fluid. The inner core of each corpuscle is connected to a single myelinated PC nerve [7]. The size of the all corpuscles is, on average, 1 mm long and can reach 4 mm in humans. There are about 350 of these corpuscles in the index finger and 800 in the palm [70]. Their size has made them the most studied mechanoreceptor.

This receptor is the most sensitive of all types. Pacinian channels can resolve skin displacement of 40 nm at their peak of sensitivity. This value falls down to 3 nm when the corpuscle is directly excited during in vitro studies [7]. They are effective in a frequency bandwidth from 40 Hz to 1 kHz and exhibit a U-shape sensitivity curve tuned to 250 Hz [65, 154, 167]. Between 60 and 250 Hz their sensitivity to vibration increases at a rate of 40 dB per decade, suggesting that the Pacinian system is sensitive to acceleration of the skin [15]. Despite their spectacular temporal resolution, their spatial acuity is low to non-existent, as PC afferents account for stimulation in a large area such as a finger phalanx [63].

The high sensibility to vibration makes them the major contributor of hand-held tools perception. When holding a probe, vibrations and shocks induced by the contact of the tip with a surface are transmitted to the hand over large skin areas. Vibration of a tool which act perpendicular to the skin are best perceived than tangential ones [17].

It is believed that Pacinian corpuscles have a key role in the perception of fine texture with the bare finger. Using the aftereffect of vibration exposure at 250 Hz to suppress PC afferent responses, Hollins and colleagues [54] observed a significant decrease in the discrimination and the perceptual scaling of fine textured sandpapers. Moreover, Gescheider et al. [42] found that a 250 Hz adaptation did not impair estimates of millimeter-scale raised cones but the shapes did feel smoother. This implies that the PC afferents are responsible for the mechanical effect created by sharp transitions of the truncated cones and the fine background roughness. Also, PC channels are more sensitive to repetitive events. In fact a single dot on a smooth surface triggers mostly RA and SA I afferents whereas a sine-wave grating of 60 nm amplitude moving under the fingerpad triggers action potentials in the PC channel [82].

2.1.4 Vibration Sensitivity

The previous paragraphs showed the importance of vibration transmission and perception in the tactual exploration of textures and how it is mediated via mechanoreceptors. The present paragraph reviews the abilities of the human somatosensory system to discriminate vibrations that differ in amplitude, frequency and waveform.

2.1.4.1 Amplitude Discrimination

Vibrotactile sensitivity shares many properties with audition. For instance, the detection threshold of a sinusoidal displacement of the skin is not identical for all stimulation frequencies. The bandwidth over which the tactile system can detect vibration ranges from 0.4 Hz to 700 Hz. The typical sensitivity curve is considered to be almost flat from 0.4 to 3 Hz with a threshold at 30 μm. The threshold decreases at a rate of −16 dB per decade (or −5 dB/oct) from 3 to 60 Hz. Sensitivity in the 60–700 Hz band follows a U-shaped curve with a minimum at 250 Hz and the smallest threshold at 0.1 μm. The left part of the U-shaped curve decreases at the rate of −40 dB per decade [15].

The Weber fraction for amplitude discrimination above the perceptual threshold is about 50 % for 10 dB SL[3] and decreases to 5 % at 40 dB SL [41]. It is worth noting that this variation of the perceptual threshold constitutes proof that Weber's law does not hold for all amplitude levels. Varying waveform and frequency did not improve the human discrimination ability. Similar Weber fractions were found for 25 Hz pure tone, 250 Hz sinusoid and broadband noise. This independence to signal properties suggests that the magnitude estimates are based on the relative energy of the signal.

Verrillo et al. [170] measured the curve for identical sensation in a 25–700 Hz frequency band and found these isocurves to scale in proportion to incremental amplitude, expressed in decibels, which means that it is logarithmic with the power of the signal. The only exception was for large amplitudes (>30 dB SL) and high frequencies (100–700 Hz) where the U-shape tend to flatten. The amplitude estimate for each frequency was well explained by a power law with a coefficient of 0.89 up to 350 Hz. The exponent increases with the frequency after 350 Hz.

Several conditions influence perceptual thresholds. The area of the contact is one of these factors. The aforementioned measurements were made with a 500 mm^2 contactor. Decreasing the area of contact caused the threshold to decrease until it reached a plateau at 30 μm for areas below 1 mm^2. The accepted explanation is that small contractor areas stimulate only the surface of the skin, whereas larger contact areas transmit vibration deeper, thus excite the Pacinian channel [168]. The presence of a surround around the vibrating probe is also a factor as it blocks wave propagation and eliminates the possibility of spatial summation [166]. The temperature also affects the discrimination. At 15 °C the threshold is three time lower than at 35 °C then it stabilizes until 40 °C [169]. Lastly, thresholds are slightly different when holding a tube [17], a pen [59] or a sphere [60], but the curves of frequency sensitivity nonetheless possess the same shapes.

2.1.4.2 Temporal and Frequency Discrimination

Mahns et al. [96] measured frequency discrimination for four frequencies (20, 50, 100 and 200 Hz) when stimulating the glabrous and the hairy skin. They found the Weber fraction $\Delta f / f$ to vary from 36 % to 14 % for reference frequencies f of 20 Hz and 200 Hz respectively, when the fingertip is stimulated. Detection thresholds are substantially higher in the hairy skin, probably because of the dynamics of the skin and the fact that Pacinian corpuscles are located deeper in the hairy skin than in the glabrous skin. Other studies have found Weber fractions as low as 6 % [45, 81] for low frequency (30–40 Hz) stimulation of the fingertip.

As in audition, differences between combinations of two pure sinusoidal signals with different phase are not always well perceived [10]. Phase difference is an important feature as two signals combined with different phase delay give rise to a

---

[3]Decibel above the sensation level.

variety of different waveforms. In the case of vibrotactile sensitivity, low frequency signals (10 Hz + 30 Hz) with four different phases (0 to 270° by 90° steps) are well discriminated. At high frequencies, however, the discrimination of phase differences in comparison of combinations of two vibrotactile stimuli (100 Hz + 300 Hz) is poor. Previous adaptation to a 10 Hz signal decreases phase discrimination abilities at high frequency, even further. An explanation given by the authors is that the PC channel integrates the energy of the signal over time, and therefore has a weaker temporal consistency. Low frequencies however are mediated by the RA channel, which encodes the complex waveforms and responds with more accurate temporal resolution.

Also, concerning temporal resolution, the somatosensory system can resolve stimuli separated by 10 ms. Below this value, both stimuli are felt as one [123].

To conclude this review of the perceptual and physiological literature on touch, it seems that during bare finger exploration of textures, the human sensory system acts in a peculiar fashion. A net friction estimate is correlated with the lateral force that shears the fingertip, as possibly mediated by SA II afferents. Fine texture, however, does not depend on spatial deformation of the finger but instead on vibrations generated by the sliding motion of the finger on the surface. The Pacinian channel is a candidate for the mediation of fine texture, despite the fact that the contribution of RA afferents is possible. Finally, roughness appears to be conveyed through rapid lateral displacement of the skin and via the normal force that the finger applies to the surface has a minor influence. Hence to reproduce fine texture for the fingertip, one should, arguably, focus on variations of lateral displacement. Figure 2.1 summarize some attributes of mechanotransduction during sliding of a finger on a textures surface.

2.2 Bio-tribology of the Skin

Mechanical waves, from which the central nervous system interprets the sensation of roughness and texture, originate from friction between the skin and the surface. Sliding motion creates acoustic energy that is radiated both in the tissues and in the surrounding environment [2]. This section explores the complex structure of the fingertip, the behavior of the skin and the relation between the production of friction and vibration during tactual exploration of texture.

2.2.1 Fingertip Anatomy

Touch sensation is present everywhere on the surface of the body thanks to the innervation of our skin. But during dextrous tasks, it is the hand, and especially the fingertips, that are the most engaged. Tactual perception is also often realized by stroking, tapping or pinching the surface with the tip of a finger. Fingertips are shaped in a specific way that provides compliance and at the same time high strength. In essence,

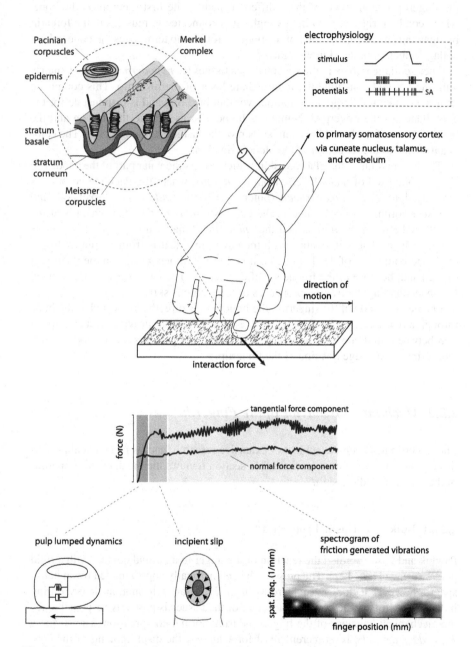

Fig. 2.1 Summary of the touch perception. Includes mechanoreceptors location (*top left*), afferents responses (*top right*), and mechanical interaction during tactual exploration (*bottom*). Inspired from [4, 65, 113, 120, 179]

the fingertip is composed of three different media. The first medium is the bone, which combine rigidity with light weight. It is connected to muscles in the forearm by ligaments fixed on each side of the bone. A rough cap terminates the bone, where collagen fibers link the skin with the tip.

These collagen fibers in the subcutaneous tissues (or *hypodermis*) create a matrix with the fat, and constitute the intermediate layer of the fingertip. This composite material can endure large deformations without breaking and is responsible for the compliance of the fingerpad. Numerous blood vessels irrigate this region and the best evidence of the blood flow can be seen when pushing of the fingertip. The color changes due to the reduction of the blood irrigation.

The last medium is the glabrous skin, which covers the volar part of the hand. The skin is composed of many layers, from the *dermis*, where the Pacinian corpuscles are lodged, to the *stratum corneum* (outer layer of the *epidermis*). Merkel cells and Meissner corpuscles are located in the curvy *stratum basale* layer which is at the boundary between the *dermis* and the *epidermis*. Finally, the *stratum corneum* is the outer layer that is in contact with the environment. It is from 10 μm to 40 μm thick and composed of dead epithelia. It is 20 to 30 times thicker on the volar part of the hand, because of the frequency of contact. Its mechanical behavior is crucial for understanding the friction properties of the glabrous skin.

On the other side of the fingertip, the fingernail is directly connected to the bone through a dense array of collagen fibers. This arrangement provides a stiff connection between the bone and the nail [138], therefore represents a good location for measuring the average position of the fingertip.

2.2.2 Mechanical Behavior of the Fingertip

The fingertip has been the subject of extensive research, and its mechanical properties continue to fascinate scientists. This section reviews the mechanical properties of the fingertip and the skin.

2.2.2.1 Bulk Viscoelastic Properties

Pawluk and Howe studied the reaction of the finger to a normal contact with a rigid planar surface [126, 127]. The pulp can be geometrically approximated by an hemisphere, however it does not adhere to linear Hertzian deformation theory during indentation. In fact, the complex structure of the tissues is probably responsible for the mechanical behavior of the fingertip during normal compression. The stiffness $K_n = dF_n/dx$ (dF_n is an increment of force and dx the displacement) of the contact linearly increases with the force. This results in an exponential form for the force-displacement curve when measured in quasi-static conditions:

$$K_n = \frac{dF_n}{dx} = aF_n + b \quad \text{and} \quad F_n = \frac{b}{a}\left(e^{(x-x_0)} - 1\right) \tag{2.1}$$

where x_0 is the initial position for which the force is null. The force is 0.3 N at 1 mm indentation and 1.8 N at 1.5 mm, which correspond to stiffnesses of 1 $\mathrm{N\,m^{-1}}$ and 4 $\mathrm{N\,m^{-1}}$ respectively. The authors also measured the distribution of pressure, and modeled it as a Gaussian curve whose center pressure reaches 30 kPa for a force of 2 N. During these indentations, the global volume was determined to be reduced by 1 %, which makes the assumption of incompressibility acceptable [148].

The pulp is also viscoelastic. One of the effect of the viscoelasticity is that increasing of the rate of pulp compression results in a stiffer reaction of the fingertip. At a indentation speed of 80 $\mathrm{m\,s^{-1}}$ the stiffness is 4 times higher than at 0.2 $\mathrm{m\,s^{-1}}$. The viscoelasticity is also visible when the pulp is rapidly indented to a fixed value. Immediately after the indentation, the force returned by the pulp relaxes across time. The relaxation response time is composed of three decays with time constants 4 ms, 70 ms and 1.4 s. This viscoelastic behavior probably protects the fingertip against shocks by stiffening the pulp and dissipating the contact energy. The relaxation function $G(t)$ is often modeled by a sum of decaying exponential. The resulting force of the viscoelastic model $P(t)$ to an arbitrary displacement is then described by the following convolution

$$P(t) = \int_{-\infty}^{\infty} G(t-\tau) \frac{\partial F_n(x(\tau))}{\partial x} \frac{\partial x(\tau)}{\partial \tau} d\tau \qquad (2.2)$$

During tapping, the finger displays a hysteresis effect that dissipates 80 % of the entry energy [136]. The non-linear viscoelastic model described by (2.2) successfully predicts the mechanical reaction resulting from voluntary tapping on flat surfaces [61].

The lateral deformation of the fingertip demonstrates similar mechanical behavior to that of normal indenting. It exhibits hysteresic losses and non-linear relaxation behavior. However the quasi-static force is linear with the lateral displacement for tangential forces up to 5 N [125]. At low frequencies, Nakazawa et al. [113] identified the fingertip in lateral deformation as a Kelvin model corresponding to a spring connected with a damper in parallel. The properties of the pulp vary with many parameters such as the normal force and also change from person to person. Stiffness values range from 0.5 $\mathrm{N\,m^{-1}}$ to 3 $\mathrm{N\,m^{-1}}$.

For the whole finger, linear lumped element models are sufficient to describe the mechanical impedance for rapid transients [48]. The parameters are a mass of 6 to 8 g, a damping in a range of 2 and 5 $\mathrm{N\,m^{-1}\,s}$ and a stiffness that evolves linearly with muscle activation from 0.2 $\mathrm{N\,m^{-1}}$ to 3 $\mathrm{N\,m^{-1}}$. The difference between the fingertip alone and the whole finger is to attributed to the joint which contributes added flexibility and greater moving mass.

2.2.2.2 Local Deformation

The high deformability of the fingertip also facilitates establishing large contact areas between the skin and the object being touched even with low forces. Pushing on a flat surface with a force of 1 N results in an area of contact 60 % as large as the value that is reached when pushing with a force of 10 N [4, 137, 144]. The friction of the skin directly depends of the area of contact between, and the large contact area enables easier control of the grip force.

The deformability of the skin is also a key factor in the large compliance of the fingertip. Measurements have shown that shearing the skin can lead to 100 % of deformation without any damage, yet the Young's modulus remains relatively high ≈ 1 MPa [172]. During lateral motion, a single bump on the surface can yield a skin stretch larger than 30 % [89].

The response to vibration seems to be bimodal. At low frequency, normal indentation with a probe is in phase with the displacement of the probe, but above 100 Hz the probe starts to decouple which implies that the tissue response is delayed by viscosity [24]. Surface mechanical waves can travel relatively long distance, at a speed of approximatively $1.6 \, \mathrm{m\,s^{-1}}$ [44]. The amplitude of the waves decays with the square of the distance of stimulation, which results in a decrease of one third at a distance of 50 mm. This may assist with sensing fine textures, as the vibrations are transduced by Pacinian corpuscles along the course of waves propagation [30].

2.2.2.3 Models

The observations of the fingertip reaction to various load described in the previous paragraph, have been the foundation of lumped parameters models of the fingertip. These models are usually valid for only one load distribution and do not extend easily to other boundary conditions. Moreover, they do not give any details about the stress distribution inside the tissues. This section reviews the continuum mechanics models that attempted to explain the diffusion of stress inside the fingertip and link the strain pattern to the spatio-temporal tactile perception.

A number of models are based on Boussinesq's equation for the deformation of an infinite plane and consider only the superficial layer of the skin. They also approximate the skin as a plane by unrolling the fingertip. With a simple elastic model it is possible to predict the firing rate of SA I afferents, which are sensitive to spatial determinant [37, 130]. The addition of viscosity in the definition of the material makes this model more realistic, enabling it to predict the spatiotemporal sensitivity of RA and PC afferents [165]. However, the linear slab model combined with Hertz contact does not correctly predict the non-linear behavior of the bulk fingertip.

Srinivasan proposed a model composed of an elastic membrane filled with an incompressible fluid. The so-called "waterbed" model predicts with a relatively good precision the indentation of the fingerpad with a thin line [149]. These models are also successful at predicting the growth of the area of contact when indenting with a flat surface [137]. The two materials in the model are emulating the role of the skin (the membrane) and the subcutaneous tissues (the fluid). Replacing the membrane by finite-element thin shells allow to accommodate more realistic geometries by still keeping a reasonable computational time [151].

Finite element models have been developed to capture the complex mechanical assembly that is the fingertip and try to unravel the mechanism of tactual perception. The preceding "waterbed" models by their constitution fail to describe the strain inside the tissues and cannot predict the firing of the mechanoreceptive afferents. Moreover, each layer of the skin can be modeled with a accurate geometry. The refine geometry makes possible the investigation of mechanical effects of papilla ridges for instance [93]. Two and three dimensional models have lead to thinks that SA I respond to the strain energy density [29, 146]. These model give a good incentive of the relationship between anatomy and neurology, but the risk of numerical error due to the non-homogeneity of the tissues can be critical for the quality of the results. Dynamic parameters of the skin and subcutaneous tissue can also be modeled and used to predict the viscoelastic behavior of the skin to various dynamic stimulations [180].

2.2.3 Friction Properties on Rough and Smooth Surfaces

Many phenomena are salient to feeling a surface by stroking it with the fingertip. Some of them are related to the viscoelastic behavior, but most of the observations are also based on the unique frictional properties of the finger and skin.

2.2.3.1 Basic Notions of Contact and Tribology

Friction is the reaction force to a motion caused by the relative motion of two surface sliding or rolling on each other. The science behind friction is called tribology, and studies the conditions for various phenomena to arise during relative motion of two surfaces in contact. Two types of friction are often described, adherence (stick) and sliding friction (slip). The former occurs when the contact is not broken and bonds tie both surfaces together. When the tangential force reaches a critical value the bonds break and objects start sliding relative to one another. A force opposite to the motion is still present, but often decreased compared to the one needed to break adherence.

The friction force F_t is proportional to the real area of contact A_r [16]. The latter can be expressed as:

$$F_t = \tau A_r \tag{2.3}$$

with τ the interfacial adhesive stress, which can depend on various parameters like the presence of a fluid between the contacting bodies or chemical properties of both materials [128]. The roughness of both surfaces is also a key factor in the resulting frictional force because it reduces the surface of the real contact area. In the contact region, higher peaks on both sides are touching and the friction force is related to the real contact area of these peaks. The consequence is that the real area of contact is smaller than the apparent area. In the case of rigid materials, the real area of contact is found by considering the plasticity of the peaks that are compressed $A_r = F_n/\sigma$ where F_n is the normal force applied between both object and σ is the yield stress limit. Amonton's law states that the friction of solid is linear in the coefficient of friction $\mu = F_t/F_n = \tau/\sigma$ and does not depend on the area of contact. This is demonstrated by combining the previous expression with (2.3).

Skin and rubber have similar behavior, sharing a low elastic modulus and a high internal dissipation. During sliding contact with a rigid surface, both behave in a similar fashions [1] and the study of elastic material provides much information about the friction of the skin. In contrast to rigid surfaces, rubber contact does not follow Amonton's law but instead it seems that Van Der Waal's forces on the molecular chains of the rubber are dominant [5]. It follows that the adhesive shear stress has a much higher value but also that the real contact area is close to the apparent value. Therefore for spherical contact, Hertz's theory links the apparent area A_a to the normal force with

$$A_a = \pi \left(\frac{3RF_n}{4E^*} \right)^{\frac{2}{3}} \tag{2.4}$$

with $E^* = \frac{E}{1-\nu^2} = \frac{4}{3}E$ and E is the Young modulus of the soft material, ν is the Poisson coefficient, often taken equal to 0.5 for rubber and skin. R is the radius of curvature of the fingertip providing that the counterbody is flat. This leads to a relationship between the normal and the friction force:

$$F_t = \tau \pi \left(\frac{9RF_n}{16E} \right)^{\frac{2}{3}}. \tag{2.5}$$

The relationship between both forces is not linear by nature even if the materials are taken to be linear. As a more general case, consider that $F_t \propto F_n^{\frac{2}{m}}$ with $2 < m < 3$. For $m = 2$ the contact is based on plasticity, whereas $m = 3$ describes a contact based only on elasticity. Also, considering the roughness of the surface as a self-affine fractal can lead to more complex models that describe the friction behavior with a better accuracy [129].

The adhesive shear stress is correlated with the molecular bond that ties both surfaces. As a consequence, its value depends of the speed of the relative motion. When there is no sliding, and the contact is static, this value is higher than when both surfaces are moving relatively. This can be explained by the fact that the bonds slowly form during static contact and strengthen the adherence. During sliding, the bonds are broken and cannot reform, leading to weaker adhesion. This theory also explains the increase in static friction force with increasing rest time.

2.2.3.2 Static Friction

Grasping and manipulation processes regulate the amount of normal force to avoid slippage, therefore keeping the tangential force below the maximum static friction force [64, 179]. Primates fingertips contact behavior is well explained by Hertz's contact law that is expressed (2.5) [174]. The authors of this study validate the model with measurements of the real area of contact and deduce that fingerprints do not likely increase friction, as they reduce the real area. In fact, the real area of contact is about 66 % of the apparent area, mostly due to the void that creates ridges.

The adhesive shear stress τ follows a linear relationship with the normal pressure p as follows:

$$\tau = \tau_0 + \alpha p \tag{2.6}$$

where τ_0 and α are coefficients [18]. In the case of glass and polypropylene on skin $\tau_0 = 4.8$ and 6.1 kPa and $\alpha = 0.8$ and 2.0 respectively [1].

The discrepancy between the two materials is explained by the surface free energy. Porous glass is hydrophilic and has low friction whereas polypropylene is hydrophobic and exhibits large friction forces. The behavior is explained by looking at moisture behavior. In the case of hydrophobic materials, water does not have any place to flow and therefore surface tension is amplified [124]. Moreover, moisture softens the *stratum corneum* and therefore the skin is more inclined to fill the contact thus increasing the friction. Production of sweat is believed to be modulated by the body to increase the friction and therefore reduce the require grip force (normal force) [3]. The friction coefficient follows an inverted U-shape curve with the moisture level, being relatively low for dry and wet skin but increases when the skin is damp [4, 161].

Small scale surface roughness (Ra $< 100\ \mu$m)[4] has an impact on the friction force when stroking the skin against hard material such as metal and plastics [50]. It has been observed that friction decreases with increasing roughness. Similar observations have been made with paper grade [140] and for fabrics [33]. These results agree with the classic tribology theory which states than a smoother surface induces more contact points therefore more friction. For rougher surfaces (Ra $> 100\ \mu$m) however, the friction tends to increase with roughness, but after a certain value the friction force reaches a plateau independently of the surface material [159]. The effect is caused by the skin, which conforms with the surface. The peaks and valleys are separated with a specific spatial period and lock onto the fingerprint ridges. At the beginning of sliding, the skin has to deform to overcome the peaks, increasing the force needed to break contact. Moreover, these protuberances also adhere with the skin, increasing at some point the area of contact [160].

2.2.3.3 Incipient Slip

The transition between stick and sliding states of a fingertip about to slip is a abrupt but not instantaneous event, where the contact is rapidly breaks from the outer ring of the contact toward the center. This event, called incipient slip, comes from the local equilibrium between the normal and the shear interfacial pressure. Hertz contact predicts a quadratic distribution of the normal pressure with maximum pressure

---

[4]Ra measurement corresponds to the average of the absolute value of the height of the asperities on the surface.

at the center and zero pressure on the outer ring. Shear stress, however, follows a hyperbolic distribution. Therefore with increasing tangential force, the shear stress will locally overcome the static friction, and the outer region will start to slide [66].

This behavior has been predicted by classical mechanics in the case of rigid solids and has also been observed in the case of the incipient slip of a finger [4, 152]. However the transition does not behave exactly like the classic Midlin–Cattaneo contact theory. This is probably because this model does not consider differences of friction coefficient between the stick and the slip state. If this difference is taken into account, it seems that the end of the transition is characterized by a minimum adhesion area above which the tangential force cannot increase [158].

2.2.3.4 Sliding Friction

Sliding friction occurs when incipient slip ends and all the contact area is in the slip state. At this stage, the finger and the object are moving tangentially relative to each another. In this state of contact, moisture is also a significant factor. On smooth non-porous surfaces, the sweat will accumulate in between the ridges leading to a increase in friction over time. The occlusion of the sweat glands is avoided with porous media, and results to a lower friction force [124]. Velocity, however, does not seem to have a direct or systematic effect on the value of the coefficient of friction.

Stroking a soft or a hard material on non-smooth surface generates vibrations that radiate, in part, as sound. For instance, noise generated by a rigid blade sliding over a rough surface exhibits a sound pressure that is proportional to the roughness estimated, raised to the power 8 to 18. The value of the exponent depends on the inclination of the blade relative to the surface [117]. Such a simple relationship cannot be drawn for finger-surface interaction, because of the complexity of the mechanism at play [2].

The low tangential stiffness of the fingertip makes it prone to stick-slip oscillations. These oscillations appear when an elastic force pulls a slider. If the static and dynamic frictions are not equal, the motion can oscillate between a stick and a slip state. This effect is often present with glass and other smooth non-porous material which exhibit a high friction with the finger [1].

2.2.4 Implication for the Perception with the Bare Finger

Touch is particularly tuned to sense the mechanical phenomena when finger is in static or dynamic contact with various shapes and surfaces.

2.2.4.1 Mechanical Illusions

Similarly to the visual system, the central nervous system does not acquire the complete strain tensors field that results to the contact, but makes simplifications. These simplifications are based on the fact that mechanoreceptors are sensitive to strain, and also that different directions of the strain at the surface can lead to the same pattern. This information loss creates mechanical tactile illusions.

For instance, if shear stress applied in a band at the center of the fingertip while the outer stays stationary, it is felt as a bump. Finite element analysis reveals that normal and tangential loads produce similar stain patterns at the presumed location of the mechanoreceptor. Since a bump is more common than a lateral stress field, the brain makes sense of the strain pattern by feeling a bump [112]. Similarly, distributed shear stress on the surface well approximates the strain distribution that produces a wave on the surface, at the location of the mechanoreceptors [72, 173].

Viscoelasticity of the tissue also have effects on tactile perception. Moy et al. showed that the relaxation of force during incremental displacement can bias the detection of edges and texture [111]. At last, during sliding of the finger, adding lubricant, decreases the friction, and therefore biases the estimation of roughness [142].

2.2.4.2 Effect of the Fingerprints

There is still much debate about the effect of the fingerprint ridges on tactual perception. It is believed that the papilla ridges, mirrors of the fingerprint ridges, act as levers to magnify the stress [22, 93]. Friction shears the ridges and therefore compresses one of the Meissner corpuscles and stretches the second. Moreover, the presence of ridges concentrates the strain locally and enhances the spatial sensitivity by limiting the cross talk of mechanical stimulus [40].

Another hypothesis looks at the regularity of the contact with the glabrous skin. In fact, the glabrous skin, and especially that found in the fingertip, is subject to mechanical loading for long periods. Ridges could act as a compliant mechanism that avoids damage and strengthens the skin [174]. This strengthening limits the creation of blisters and sores. The finding that the skin is more elastic along the ridges than across them supports the hypothesis that the skin behaves like a deforming bellow [172].

Recently, there has been a debate about the importance of fingerprints in texture perception. Scheibert et al. [135] observed, by means of an artificial finger, that ridges on the surface enhance vibratory signals during perception of fine texture by exciting the frequency where PC afferents are the most sensible. However, this hypothesis was challenged by other researchers, as human fingertips are more complex that simple linear elastic materials [27], and because the presence of fingerprints does not affect signal processing when using finite element analysis [93].

To conclude this overview of the mechanical phenomena that are involved in finger-surface interaction during tactile exploration, it could be observed that the finger is a complex mechanical assembly that adapts to a large variety of shapes and regulates its friction by sweating to optimize grip. Part of the coarse elements of a surface are perceived by the strain field induce by the skin on the shape. The other part of the information is based on the friction noise that is generated during sliding of the finger. Friction noise is composed of vibrations that propagate in the tissues. Fine texture is inferred from these vibrations, and the spatial distribution of strain is less important. Figure 2.1 summarize the main mechanical events that occurs during sliding of a fingertip on a textured surface.

2.3 Reproduction of Impact and Slip

This section describes a number of devices devoted to the reproduction of tactile sensations. Such devices were motivated by research in diverse areas such as teleoperation, robotic sensing, haptic feedback in virtual reality, and consumer products. The focus is put on those technologies that are able to capture or reproduce sensations of roughness, impact and slip. Interfaces based on pin matrices intended to deform or stretch the skin or surface displays that move a surface under the skin to induce shape sensation will not be discussed in this dissertation. The reader can consult other extensive surveys for additional information on these approaches [8, 122].

2.3.1 Signal Acquisition

Researchers have attempted to capture the complexity of the mechanical interactions arising during the exploration of surfaces, either with a probe or with a bare finger. Some studies have been devoted to recording these interactions for haptic simulation or texture perception.

2.3.1.1 Haptic Interaction Recording

In order to record haptic interaction, many projects propose to acquire the high frequency acceleration that is transmitted to the tip of a probe during exploration. A probe sliding on a rough surface is solicited by surface stresses that change through time as the probe moves. The variation of surface stresses induces mechanical vibrations in the probe that can be picked up by accelerometers. With this simple method, roughness can be sensed, recorded, and modeled [83, 118, 119].

In the case of bare finger exploration, sensing interaction forces or skin displacements is more challenging. Ideally the interaction should be measured at the interface between the finger and the surface, but the addition of any type of sensor would necessarily disturb the interaction. It is also possible to measure the displacement of tissues in the finger, for instance by means of a Hall-effect sensor [9], acceleration of the nail [74, 100] or the noise radiated by the finger [155]. These direct measurement techniques employ sensors that are attached to the skin with the following shortcomings. These sensors have a finite mass which may not be negligible compared the impedance of the tissue to which they are attached, perturbing the measurement. But the hardest obstacle is to establish a reliable reconstruction of the skin vibration from distal measurements.

2.3.1.2 Biomimetic Robotics Fingers

Artificial hand research has been focusing on grasping and manipulation. In this context, the tactile sensors, if any, that they include in their design are mostly intended to detect contact and slip. Many are based on an array of sensors that pick up spatially distributed, low frequency interaction in the expectation to detect the shape of the touched object. They are often not well suited for rapid interaction and are sensitive to rapid mechanical transient signals. For recent reviews of these distributed sensors and their fabrication techniques please consult [28, 186].

However, some of the sensors are based on the observation of our sensory system. For instance, Howe et Cutkosky proposed a piezoelectric sensor that reacts to the rate of change of the strain inside a rubber band. This approach mimic the role of Pacinian and authors showed that they could detect slippage and even texture [56, 57]. Robots adjust their grip force accordingly to the captured signal of incipient slip. As for humans, they use the relative tangential motion for the artificial fingers to measure and classify natural textures from the friction-generated vibration [73]. They noted that measuring acceleration rather than displacement is easier for high frequency measurements, as the sensitivity of accelerometer to displacement amplitude evolves with the square of the frequency. Reciprocally, for low frequency and static measurement, displacement gives more signal than acceleration. In the spirit of biomimetism, some artificial fingers are designed to sense the distribution of stick and slip region of the contact and use this information to estimate the ratio of normal and friction force [94, 177]. Fine control and manipulation is achieved by tactually sensing the contact state with the hand-held object.

2.3.2 Reproduction by Force Feedback

To render tactile and kinesthetic sensations, force feedback devices are often used. These devices amount to robotic mechanisms that apply forces relatively to the mechanical ground in response to the displacement of the finger or the hand.

The reproduction of tactual texture with force feedback has been initiated by Minsky during the development of the *Sandpaper* system [105]. They used a two degree-of-freedom (DOF) joystick to synthesize textures. The lateral forces were modulated as a function of the position and of the gradient of the virtual height field. This method produced a sensation on 2D plane that could be associated with roughness [104].

Other algorithms were proposed to extend the rendering of roughness to 3DOF. They used either a oscillating force that is normal, tangential or in both directions to the virtual non-textured surface. Normal forces approaches tend to make the textures feel 'frictionless'. Texture algorithms themselves can lead to unwanted artifacts and oscillations [21].

The manipulandum and its control is also a critical element for the reproduction with high fidelity. Campion and Hayward used Nyquist and Courant–Friedrichs–Lewy conditions to draw the fundamental limits in terms of force, position and temporal resolution for correct reproduction [20]. For fine texture reproduction, the resolution of encoder should be as low as 1 μm and the force should be modulated by milli-newtons increments to match human perception thresholds. They also explored the effect of the bandwidth of force feedback and concluded that a device capable of stimulating the finger from DC to 700 Hz with guaranteed stability was difficult to achieve. Commercially available haptic devices, such as the Phantom, are limited by their first structural modes, in this case at 30 Hz.

To overcome the bandwidth limitation several approaches are possible. Transient and rapid mechanical events can be reproduced by vibrotactile shakers mounted on the thimble of force feedback devices [58]. Another approach is to couple two motors of different sizes. The small motor is fast but does not deliver large forces whereas the larger motor produces more torque at the expense of a larger inertia [103]. Another solution is to compensate the narrow bandwidth of a haptic display with an inverse filter. The inverse filter however, should depend on the configuration of the device, which makes the implementation complex [79].

Ordinary force-feedback devices in general suffer from the complexities of their mechanical properties and are prone to introduce instabilities and artifacts.

2.3.3 Controlled Friction Displays Approaches

Grounded, electric powered, force-feedback devices are not the only approach to rendering textures. An alternative approach consists of modulating the friction between a surface and the fingertip. This family of devices is increasingly gaining attention because of their compatibility with touchscreen interfaces and the simplicity of their open-loop control.

It is known that surface acoustic waves propagating in rigid substrate can reduce the coefficient of friction between the substrate and an object. Such effect was put to practice by placing an aluminum foil between the fingertip and the surface, eliciting a sensation of reduced stickiness (static reduction) and even roughness (alternative forces) [153]. Electrostatic forces can also be used to change the frictional forces by alternatively attracting and repulsing, regardless of the direction of motion [181].

Electrostatic attraction can also be achieved directly on the fingertip, eliminating the need for an interposed slider. The range of force is greatly reduced and low frequencies cannot be felt. However, if a alternative current of frequency greater than 100 Hz is applied to the electrodes, a vibrotactile sensation is induced when the finger moves [156]. Recently, this technology have been implemented by TeslaTouch [6] and Senseg [97] with the goal to be deployed in multi-touch devices.

Bare finger friction can also be modulated by the squeeze-film effect that exploits the thermodynamic non-linearity of gases. When exciting a plate with normal ultrasonic vibrations, a film of air is pumped under the finger, reducing the friction with the plate. Air pumping was first applied to tactile rendering by Watanabe and Fukui [175]. Two groups, Biet et al. [12] and Windfield et al. [178], later presented devices that uses this phenomenon almost simultaneously. The first group used a standing wave in a berylium-copper plate to pump the air below the finger and the other used the first bending mode of a disc piezoelectric transducer to achieve ultrasonic levitation. Improvements have been made such as the use of multiple bending normal mode of a plate to extend applicability to large displays [99].

2.3.4 Vibration Based Displays

Vibrotactile devices do not produce low-frequency forces like force-feedback devices but rather focus on the fact that human perceptual system is most sensitive to high-frequency signals. Exciting the skin with vibrations is an efficient way to elicit tactile sensations.

2.3.4.1 Eccentric Mass Motors

The most common stimulators are the 'rumble motors' that are found in almost all cellphones and game controller. Their low price and high efficiency make them the technology of choice for alarm-type tactile signals. The stimulator is a DC motor with an eccentric mass attached to the shaft. Activation creates rotating radial forces due to the eccentricity. Despite its simplicity and efficiency, this technology is ill-suited for generating rich sensations since signal amplitude and the frequency are inherently coupled from the principle of operation. They also suffer from poor temporal resolution.

2.3.4.2 Electromagnetic Shakers and Recoil Motors

Richer vibrotactile sensations are made possible by the inertial motor based on voice-coil transducers or on piezoelectric actuators. They can be applied to vibrate the screen or the enclosure of a device to produce tactile sensations [39]. It can also or be used to directly stimulate the skin [109, 184]. Voice-coil actuators use Laplace forces to move a magnet suspended in the housing by a membrane. The magnet serves as the inertial slug and moves over a single degree of freedom. The moving slug, driven above the resonant frequency, creates a force that pushes back by conservation of momentum on the housing and therefore on the skin. These recoil motors can reproduce acceleration over a large bandwidth (typically 50 Hz to several kHz) being limited in the low frequency by the suspension elasticity, maximum displacement, and the mass of the slug. In the high frequencies, the structural modes determine the limit of the response. It is advantageous to design-in a low coefficient of quality to damp the resonance and extend the flat bandwidth.

Recoil motors enable fine control over of vibrotactile signal without the need of a ground reference. They have been used in many demonstrations. Yao et al. described a stick equipped with such voice-coil motor to simulate a virtual rolling ball

[183]. Textural sensations can also be created with pen-based interfaces by inducing vibrations in the stylus [133]. Larger motors can be used to stimulate the foot, and give the sensation of walking on virtual floor such as sand or ice [171].

2.3.4.3 Linear Resonant Actuators

Some application use linear-resonant actuators (LRA) to create a tactile sensation by amplitude modulation of a vibration at a resonant frequency. LRA have the same basic construction a recoil motors, but their coefficient of quality is much higher and they are usually less powerful. As a consequence only the resonant frequency can be felt [34]. Complex signals are usually generated by amplitude modulation. To increase efficiency, the inertial slug can intentionally be designed to impact the enclosure, creating a strong, but hard-to-control, acceleration pattern [182].

2.3.4.4 Piezoelectric Devices

Many researchers attempted to replace electromagnetic drives by other actuation means such as shape memory alloys, pneumatic, electroactive polymers and so on. These novel actuation approaches rarely reach sufficient maturity to be attractive for the consumer electronic market, due to various limitations arising from high drive voltages, manufacturing difficulties or lack of reliability. Piezoelectric transducers on the other hand can be exceedingly practical.

Piezoelectricity is the property of crystals to displace charges in a material when it is strained. Conversely, electrostatic fields applied across the material induce mechanical deformations. Owing to the large number of application of these transducers, ceramics have been developed to enhance the achievable strains. A common material is the lead-zirconate-titanate ceramic, also named PZT, which can deform to a maximum strain of about 0.1 %.

Stacking, and other mechanical arrangements of thin PZT plates, are often necessary to achieve a workable stroke displacement for a given actuator size. Piezoelectric actuators are often used in the form of bending bimorphs, which are composed of two plates bonded together and actuated in opposition. The compression of one side and the stretching of the other causes the blade to bend, achieving displacements that can reach several millimeters under reasonable voltages.

These piezoelectric bimorphs are, for instance, responsible of the motion of the pins in Braille cells [162]. They are also used to move an inertial slug inside an enclosure, or directly vibrate the screen of a portable device[132]. Their natural stiffness and low damping makes them efficient when used at resonance.

Similarly to electromagnetic LRA's, tactile signals can be generated by amplitude-modulation. Maeno et al. applied this idea to reproduce virtual textures by using an ultrasonic motor modulated with the amplitude of the measurement made by a sliding probe [95].

2.3.5 Simulated Material and Textures

Actuators are driven accordingly to the user's input. In force-feedback devices, a force is specified in response to a position input (the impedance approach). Another approach is to regulate the velocity of a handle according to an input force (the admittance approach). This section reviews the various algorithms that synthesize realistic tactile sensation accordingly to a virtual representation of the world. Most algorithms are based on the assumption that a contact can be approximated by the behavior of a single point impacting or sliding in the virtual world.

2.3.5.1 Impact

Impact transients and high frequency oscillations add realism to haptic simulations in virtual environments. Okamura et al. [116] proposed a methodology to acquire oscillations generated by the impact with real materials. The measured response was modeled by decaying sinusoids which coefficients that best fit the transient. Hard material exhibit higher frequency and lower decay rate, whereas soft material like rubber damp the collision. During the simulation, transients are replayed by adding the decaying sinusoid with amplitude proportional to the impact velocity to the amplitude of the force signal. Instead of using only one damped sinusoid, a sum of decaying sinusoid can be used to model the modal response of more complex object to an impact [163]. The impact location can also affect the vibration signature. For instance, a beam produces modal vibrations that depend on the location of an impact. Reproducing this effects is sufficient to give to the user the sensation of the impact location [145].

2.3.5.2 Roughness

Virtual textures synthesis tools rely on physical properties of real textures. Siira and Pai noticed that many textured surfaces were characterized by a height Gaussian distribution. As a result, they proposed a stochastic model to synthesize textures by addition of a random component to the rendering of a smooth surface. They took advantage of the fact that normal distribution is independent of resampling and that the norm and standard deviation did not depend on velocity. This observation led to a simplified computation able to generate white surface noise at every increment of time. The simplicity of the rendering in the temporal domain has the disadvantage that the spatial consistency is not maintained. As a result, the same point in space could be rendered with different values [139].

The stochastic model was extended to render other features of textures. Bump and deterministic geometries can be modeled by their Fourier decomposition and the stochastic behavior described by a sum of Gaussian distributions. Other random processes such as band-limited noise or banks of filtered noise can be used to replace the Gaussian white noise distribution [38]. A texture model described in the spatial domain was then interpolated in the temporal domain for rendering. Perceptually, it was found that the roughness sensations were correlated with the standard deviation of the Gaussian distribution.

The idea of space-based description has been used by van den Doen et al. [164] to render the sounds produced by stoking a stick on a surface at audio rates. To reproduce the recording, they used a "phonograph needle" model. Samples were recorded at constant speed, v_{ref}, and replayed at a rate of v/v_{ref} by interpolating the missing samples. Moreover, the audio volume was scaled by the square root of the power dissipated by friction. If friction is assumed to follow Amonton's law of friction, then the acoustic signal is $a(t) \propto \sqrt{\mu |F_n v|}$ where μ is the coefficient of dynamic friction, F_n the normal force, and v the velocity.

Surface topology can also be described by fractal noise. Fractals are irregular self-similar structures often found in nature processes [98]. Costa et al. [26] computed the height profile of a textured surface by synthesizing the signal as a Fourier series which coefficients generated by $1/f^\beta$ power spectral noise density. Experiment indicated that the roughness estimates were strongly correlated with the root-mean-square of the height profile. Different fractal dimensions evoked different textures.

2.3.5.3 Friction Related Events

Other phenomena, such as stick-slip oscillations, have been virtually rendered. Stick-slip oscillations occur when the friction force is higher during the stuck state than during the slip state and when there is a forcing term. When these conditions are met, a system can oscillates between the stuck state and the slip state according to the mean velocity. Konyo et al. proposed to trigger a 250 Hz decaying sinusoids produced by a piezoelectric transducer at the transitions to provide a simulated feeling of incipient slip under the bare finger [77].

2.4 Conclusion

Texture is an important attribute of the objects; it has therefore received a commensurate amount of attention in the literature. The amount of realism of virtual environment depends greatly on the fidelity of the visual, audio, and haptic texture models. Much work has been dedicated to the understanding of perception of texture through direct contact with a bare finger. The consensus is that the sensation of roughness plays a central role in the perception of textures and that roughness sensations are mediated by rapid mechanical events rather than spatially coded information.

To date, research regarding the encoding of tactual roughness encoding and its artificial reproduction has been mainly focused on the rigid-probe model. This focus is easy to explain since recording the vibrations of a probe stroking a surface only requires an accelerometer and reproduction can be accomplished with force-feedback devices and/or vibrotactile transducers. In the case of a bare finger the contact condition are fundamentally different from that of a rigid probe. The analysis and the modeling of the mechanics of interaction no longer can be reduced to the movements of a rigid body. It is necessary to account for the mechanical properties of a fingertip and for skin tribology to gain an appreciation of the complexity of the interaction.

Research in biomechanics and skin tribology was to date mostly motivated by issues in motor control (grip), as well as by health and cosmetics research. As a result, the complex interactions arising during sliding on a surface have been by-and-large ignored. For instance, most studies assume quasi-static conditions or very low frequencies. Psychophysics of touch, in contrast, reveals that tactile perception operates within a range that reaches DC to 700 Hz. Moreover, very little work has been devoted to the characterization of vibrations generated when stroking a finger on surfaces.

If this challenges were solved it would be possible to transfer the knowledge of physical interactions to the rendering more realistic artificial sensations. The enhancement of the rendering techniques could be the beginning of high fidelity tactile devices and can possibly spread to several domains of applications. Transducer design would also be improved and simplified by the knowledge of the mechanics of touch.

References

1. Adams MJ, Briscoe BJ, Johnson SA (2007) Friction and lubrication of human skin. Tribol Lett 26(3):239–253
2. Akay A (2002) Acoustics of friction. J Acoust Soc Am 111:1525
3. André T, Lefevre P, Thonnard JL (2010) Fingertip moisture is optimally modulated during object manipulation. J Neurophysiol 103(1):402–408
4. André T, Lévesque V, Hayward V, Lefèvre P, Thonnard JL (2011) Effect of skin hydration on the dynamics of fingertip gripping contact. J R Soc Interface
5. Barquins M (1993) Friction and wear of rubber-like materials. Wear 160(1):1–11
6. Bau O, Poupyrev I, Israr A, Harrison C (2010) TeslaTouch: electrovibration for touch surfaces. In: Proceedings of the 23rd annual ACM symposium on user interface software and technology, ACM, New York, pp 283–292
7. Bell J, Bolanowski S, Holmes MH (1994) The structure and function of pacinian corpuscles: a review. Prog Neurobiol 42(1):79–128
8. Benali-Khoudja M, Hafez M, Alexandre JM, Kheddar A (2004) Tactile interfaces: a state-of-the-art survey. In: Int symposium on robotics
9. Bensmaïa SJ, Hollins M (2003) The vibrations of texture. Somatosens Motor Res 20(1):33–43
10. Bensmaïa SJ, Hollins M (2000) Complex tactile waveform discrimination. J Acoust Soc Am 108(3):1236–1245
11. Bergmann-Tiest WM, Kappers AML (2006) Analysis of haptic perception of materials by multidimensional scaling and physical measurements of roughness and compressibility. Acta Psychol 121(1):1–20
12. Biet M, Giraud F, Lemaire-Semail B (2007) Squeeze film effect for the design of an ultrasonic tactile plate. IEEE Trans Ultrason Ferroelectr Freq Control 54(12):2678–2688

13. Birznieks I, Macefield VG, Westling G, Johansson RS (2009) Slowly adapting mechanore-ceptors in the borders of the human fingernail encode fingertip forces. J Neurosci 29(29):9370

14. Blake DT, Johnson KO, Hsiao SS (1997) Monkey cutaneous SA I and Ra responses to raised and depressed scanned patterns: effects of width, height, orientation, and a raised surround. J Neurophysiol 78(5):2503

15. Bolanowski SJ, Gescheider GA, Verrillo RT, Checkosky CM (1988) Four channels mediate the mechanical aspects of touch. J Acoust Soc Am 84(5):1680–1684

16. Bowden FP, Tabor D (1939) The area of contact between stationary and between moving surfaces. Proc R Soc Lond Ser A, Math Phys Sci, 391–413

17. Brisben AJ, Hsiao SS, Johnson KO (1999) Detection of vibration transmitted through an object grasped in the hand. J Neurophysiol 81(4):1548

18. Briscoe BJ, Tabor D (1975) The effect of pressure on the frictional properties of polymers. Wear 34(1):29–38

19. Cadoret G, Smith AM (1996) Friction, not texture, dictates grip forces used during object manipulation. J Neurophysiol 75(5):1963

20. Campion G, Hayward V (2005) Fundamental limits in the rendering of virtual haptic textures. In: Proceedings of the first joint eurohaptics conference and symposium on haptic interfaces for virtual environment and teleoperator systems, pp 263–270

21. Campion G, Hayward V (2008) On the synthesis of haptic textures. IEEE Trans Robot 24(3):527–536

22. Cauna N (1954) Nature and functions of the papillary ridges of the digital skin. Anat Rec 119(4):449–468

23. Chambers MR, Andres KH, Duering M, Iggo A (1972) The structure and function of the slowly adapting type II mechanoreceptor in hairy skin. Exp Physiol 57(4):417

24. Cohen JC, Makous JC, Bolanowski SJ (1999) Under which conditions do the skin and probe decouple during sinusoidal vibrations? Exp Brain Res 129:211–217

25. Connor CE, Johnson KO (1992) Neural coding of tactile texture: comparison of spatial and temporal mechanisms for roughness perception. J Neurosci 12(9):3414

26. Costa MA, Cutkosky MR (2000) Roughness perception of haptically displayed fractal surfaces. In: Proceedings of the ASME dynamic systems and control division, vol 69, pp 1073–1079

27. Dahiya RS, Gori M (2010) Probing with and into fingerprints. J Neurophysiol 104(1):1

28. Dahiya RS, Metta G, Valle M, Sandini G (2010) Tactile sensing—from humans to humanoids. IEEE Trans Robot 26(1):1–20

29. Dandekar K, Raju BI, Srinivasan MA (2003) 3-d finite-element models of human and monkey fingertips to investigate the mechanics of tactile sense. J Biomech Eng 125:682

30. Delhaye B, Hayward V, Lefevre P, Thonnard JL (2010) Textural vibrations in the forearm during tactile exploration. Poster 782.11. In: Annual meeting of the society for neuroscience

31. Delmas P, Hao J, Rodat-Despoix L (2011) Molecular mechanisms of mechanotransduction in mammalian sensory neurons. Nat Rev Neurosci 12(3):139–153

32. Dépeault A, Meftah EM, Chapman CE (2008) Tactile speed scaling: contributions of time and space. J Neurophysiol 99(3):1422

33. Derler S, Schrade U, Gerhardt LC (2007) Tribology of human skin and mechanical skin equivalents in contact with textiles. Wear 263(7–12):1112–1116

34. Do Kweon S, Park IIO, Son YH, Choi J, Oh HY (2008) Linear vibration motor using resonance frequency. Google patents. US patent 7,358,633

35. Ekman G, Hosman J, Lindstrom B (1965) Roughness, smoothness, and preference: a study of quantitative relations in individual subjects. J Exp Psychol 70(1):18

36. Essick GK, Franzen O, Whitsel BL (1988) Discrimination and scaling of velocity of stimulus motion across the skin. Somatosens Motor Res 6(1):21–40

37. Fearing RS, Hollerbach JM (1985) Basic solid mechanics for tactile sensing. Int J Robot Res 4(3):40

38. Fritz JP, Barner KE (1996) Stochastic models for haptic texture. In: Proceedings of SPIE's international symposium on intelligent systems and advanced manufacturing—telemanipulator and telepresence technologies III, pp 34–44

39. Fukumoto M, Sugimura T (2001) Active click: tactile feedback for touch panels. In: CHI'01 extended abstracts on human factors in computing systems, ACM, New York, pp 121–122

40. Gerling GJ, Thomas GW (2005) The effect of fingertip microstructures on tactile edge perception. In: Proceedings of the first joint eurohaptics conference and symposium on haptic interfaces for virtual environment and teleoperator systems, IEEE Computer Society, Los Alamitos, pp 63–72

41. Gescheider GA, Bolanowski Jr SJ, Verrillo RT, Arpajian DJ, Ryan TF (1990) Vibrotactile intensity discrimination measured by three methods. J Acoust Soc Am 87:330

42. Gescheider GA, Bolanowski SJ, Greenfield TC, Brunette KE (2005) Perception of the tactile texture of raised-dot patterns: a multidimensional analysis. Somatosens Motor Res 22(3):127–140

43. Gibson JJ (1962) Observations on active touch. Psychol Rev 69(6):477

44. von Gierke HE, Oestreicher HL, Franke EK, Parrack HO, von Wittern WW (1952) Physics of vibrations in living tissues. J Appl Physiol 4(12):886

45. Goff GD (1967) Differential discrimination of frequency of cutaneous mechanical vibration. J Exp Psychol 74(2):294–299

46. Goodwin AW, Morley JW (1987) Sinusoidal movement of a grating across the monkey's fingerpad: representation of grating and movement features in afferent fiber responses. J Neurosci 7(7):2168

47. Goodwin AW, Macefield VG, Bisley JW (1997) Encoding of object curvature by tactile afferents from human fingers. J Neurophysiol 78(6):2881

48. Hajian AZ, Howe RD (1997) Identification of the mechanical impedance at the human finger tip. J Biomech Eng 119:109

49. Halata Z, Grim M, Bauman KI (2003) Friedrich Sigmund Merkel and his "Merkel cell", morphology, development, and physiology: review and new results. Anat Rec, Part A Discov Mol Cell Evol Biol 271(1):225–239

50. Hendriks C, Franklin S (2010) Influence of surface roughness, material and climate conditions on the friction of human skin. Tribol Lett 37:361–373

51. Hollins M, Risner SR (2000) Evidence for the duplex theory of tactile texture perception. Atten Percept Psychophys 62(4):695–705

52. Hollins M, Faldowski R, Rao S, Young F (1993) Perceptual dimensions of tactile surface texture: a multidimensional scaling analysis. Percept Psychophys 54(6):697–705

53. Hollins M, Bensmaïa SJ, Karlof K, Young F (2000) Individual differences in perceptual space for tactile textures: evidence from multidimensional scaling. Atten Percept Psychophys 62:1534–1544

54. Hollins M, Bensmaïa SJ, Washburn S (2001) Vibrotactile adaptation impairs discrimination of fine, but not coarse, textures. Somatosens Motor Res 18(4):253–262

55. Hollins M, Bensmaïa SJ, Roy EA (2002) Vibrotaction and texture perception. Behav Brain Res 135(1–2):51–56

56. Howe RD, Cutkosky MR (1989) Sensing skin acceleration for slip and texture perception. In: Proceedings of the 1989 IEEE international conference on robotics and automation, pp 145–150

57. Howe RD, Cutkosky MR (1993) Dynamic tactile sensing: perception of fine surface features with stress rate sensing. IEEE Trans Robot Autom 9(2):140–151

58. Howe RD, Kontarinis D (1994) High-frequency force information in teleoperated manipulation. Exp Robot III:341–352

59. Israr A, Choi S, Tan HZ (2006) Detection threshold and mechanical impedance of the hand in a pen-hold posture. In: 2006 IEEE/RSJ international conference on intelligent robots and systems, IEEE, New York, pp 472–477

60. Israr A, Choi S, Tan HZ (2007) Mechanical impedance of the hand holding a spherical tool at threshold and suprathreshold stimulation levels. In: World haptics 2007, IEEE, New York, pp 56–60

61. Jindrich DL, Zhou Y, Becker T, Dennerlein JT (2003) Non-linear viscoelastic models predict fingertip pulp force-displacement characteristics during voluntary tapping. J Biomech 36(4):497–503

62. Johansson RS, LaMotte RH (1983) Tactile detection thresholds for a single asperity on an otherwise smooth surface. Somatosens Motor Res 1(1):21–31

63. Johansson RS, Vallbo AB (1979) Tactile sensibility in the human hand: relative and absolute densities of four types of mechanoreceptive units in glabrous skin. J Physiol 286(1):283

64. Johansson RS, Westling G (1984) Roles of glabrous skin receptors and sensorimotor memory in automatic control of precision grip when lifting rougher or more slippery objects. Exp Brain Res 56(3):550–564

65. Johansson RS, Landström U, Lundström R (1982) Responses of mechanoreceptive afferent units in the glabrous skin of the human hand to sinusoidal skin displacements. Brain Res 244(1):17–25

66. Johnson KL (1987) Contact mechanics. Cambridge University Press, Cambridge

67. Johnson KO (2001) The roles and functions of cutaneous mechanoreceptors. Curr Opin Neurobiol 11(4):455–461

68. Johnson KO, Hsiao SS (1992) Neural mechanisms of tactual form and texture perception. Annu Rev Neurosci 15(1):227–250

69. Johnson KO, Phillips JR (1981) Tactile spatial resolution. I. Two-point discrimination, gap detection, grating resolution, and letter recognition. J Neurophysiol 46(6):1177

70. Johnson KO, Yoshioka T, Vega-Bermudez F (2000) Tactile functions of mechanoreceptive afferents innervating the hand. J Clin Neurophysiol 17(6):539

71. Katz D (1925) The world of touch. Original work published in Erlbaum, Hillsdale, NJ

72. Kikuuwe R, Sano A, Mochiyama H, Takesue N, Fujimoto H (2005) Enhancing haptic detection of surface undulation. ACM Trans Appl Percept 2(1):46–67

73. Kim SH, Engel J, Liu C, Jones DL (2005) Texture classification using a polymer-based mems tactile sensor. J Micromech Microeng 15:912

74. Kim YS, Kesavadas T (2006) Material property recognition by active tapping for fingertip digitizing. In: 14th symposium on haptic interfaces for virtual environment and teleoperator systems, IEEE, New York, pp 133–139

75. Klatzky RL, Lederman SJ (1999) Tactile roughness perception with a rigid link interposed between skin and surface. Percept Psychophys 61(4):591–607

76. Knibestöl M, Vallbo ÅB (1980) Intensity of sensation related to activity of slowly adapting mechanoreceptive units in the human hand. J Physiol 300(1):251

77. Konyo M, Yamada H, Okamoto S, Tadokoro S (2008) Alternative display of friction represented by tactile stimulation without tangential force. In: Haptics: perception, devices and scenarios, pp 619–629

78. Krueger LE (1982) Tactual perception in historical perspective: David Katz's world of touch. In: Schiff W, Foulke E (eds) Tactual perception; a sourcebook. Cambridge University Press, Cambridge, pp 1–55

79. Kuchenbecker KJ, Niemeyer G (2006) Improving telerobotic touch via high-frequency acceleration matching. In: Proceedings of 2006 IEEE international conference on robotics and automation, IEEE, New York, pp 3893–3898

80. Lamb GD (1983) Tactile discrimination of textured surfaces: psychophysical performance measurements in humans. J Physiol 338(1):551

81. LaMotte RH, Mountcastle VB (1975) Capacities of humans and monkeys to discriminate vibratory stimuli of different frequency and amplitude: a correlation between neural events and psychological measurements. J Neurophysiol 38(3):539

82. LaMotte RH, Srinivasan MA (1991) Surface microgeometry: tactile perception and neural encoding. In: Wenner–Gren international symposium series, pp 49–58

83. Lang J, Andrews S (2011) Measurement-based modeling of contact forces and textures for haptic rendering. IEEE Trans Vis Comput Graph 17(3):380–391

84. Lederman SJ, Klatzky RL (1987) Hand movements: a window into haptic object recognition. Cogn Psychol 19(3):342–368

85. Lederman SJ, Klatzky RL (1997) Designing haptic interfaces for teleoperational and virtual environments: should spatially distributed forces be displayed to the fingertip. In: Proc of the ASME dynamic systems and control division, symposium on haptic interfaces, DSC, vol 59

86. Lederman SJ, Taylor MM (1972) Fingertip force, surface geometry, and the perception of roughness by active touch. Percept Psychophys 12(5):401–408

87. Lederman SJ, Loomis JM, Williams DA (1982) The role of vibration in the tactual perception of roughness. Atten Percept Psychophys 32(2):109–116

88. Lederman SJ, Klatzky RL, Hamilton CL, Ramsay GI (1999) Perceiving roughness via a rigid probe: psychophysical effects of exploration speed and mode of touch. Haptics-e 1(1) (http://www.haptics-e.org). doi:10.1.1.42.4553

89. Levesque V, Hayward V (2003) Experimental evidence of lateral skin strain during tactile exploration. In: Proc Eurohaptics 2003, pp 261–275

90. Libouton X, Barbier O, Plaghki L, Thonnard JL (2010) Tactile roughness discrimination threshold is unrelated to tactile spatial acuity. Behav Brain Res 208:473–478

91. Louw S, Kappers AML, Koenderink JJ (2000) Haptic detection thresholds of Gaussian profiles over the whole range of spatial scales. Exp Brain Res 132(3):369–374

92. Lundström RJ (1986) Responses of mechanoreceptive afferent units in the glabrous skin of the human hand to vibration. Scand J Work, Environ. & Health 12(4 Spec No):413

93. Maeno T, Kobayashi K, Yamazaki N (1998) Relationship between the structure of human finger tissue and the location of tactile receptors. JSME Int J Ser C, Dyn Control Robot Des Manuf 41(1):94–100

94. Maeno T, Hiromitsu S, Kawai T (2000) Control of grasping force by detecting stick/slip distribution at the curved surface of an elastic finger. In: IEEE international conference on robotics and automation, 2000, vol 4, IEEE, New York, pp 3895–3900

95. Maeno T, Otokawa K, Konyo M (2006) Tactile display of surface texture by use of amplitude modulation of ultrasonic vibration. In: Proceedings of the IEEE ultrasonics symposium, pp 62–65

96. Mahns DA, Perkins NM, Sahai V, Robinson L, Rowe MJ (2005) Vibrotactile frequency discrimination in human hairy skin. J Neurophysiol 95(3):1442–1450

97. Makinen V, Linjama J, Gulzar Z (2010) Tactile stimulation apparatus having a composite section comprising a semiconducting material. Google patents. US patent App. 12/900,305

98. Mandelbrot BB (1982) The fractal geometry of nature. Freeman, New York

99. Marchuk ND, Colgate JE, Peshkin MA (2010) Friction measurements on a large area tpad. In: Haptics symposium, 2010, IEEE, New York, pp 317–320

100. Martinot F, Houzefa A, Biet M, Chaillou C (2006) Mechanical responses of the fingerpad and distal phalanx to friction of a grooved surface: effect of the contact angle. In: Proceedings of the IEEE conference on virtual reality, p 99

101. Meenes M, Zigler MJ (1923) An experimental study of the perceptions roughness and smoothness. Am J Psychol, 542–549

102. Meftah EM, Belingard L, Chapman CE (2000) Relative effects of the spatial and temporal characteristics of scanned surfaces on human perception of tactile roughness using passive touch. Exp Brain Res 132(3):351–361

103. Millet G, Haliyo S, Regnier S, Hayward V (2009) The ultimate haptic device: first step. In: IEEE world haptics conference 2009, pp 273–278

104. Minsky M, Lederman SJ (1996) Simulated haptic textures: roughness. In: Proceedings of the ASME dynamic systems and control division, vol 58, pp 421–426

105. Minsky M, Ming O, Steele O, Brooks Jr FP, Behensky M (1990) Feeling and seeing: issues in force display. In: Proceedings of the 1990 symposium on interactive 3D graphics, ACM, New York, pp 235–241

106. Miyaoka T, Mano T, Ohka M (1999) Mechanisms of fine-surface-texture discrimination in human tactile sensation. J Acoust Soc Am 105:2485
107. Monzée J, Lamarre Y, Smith AM (2003) The effects of digital anesthesia on force control using a precision grip. J Neurophysiol 89(2):672
108. Morley JW, Goodwin AW, Darian-Smith I (1983) Tactile discrimination of gratings. Exp Brain Res 49(2):291–299
109. Mortimer BJ, Zets GA, Cholewiak RW (2007) Vibrotactile transduction and transducers. J Acoust Soc Am 121(5 Pt 1):2970
110. Mountcastle VB, LaMotte RH, Carli G (1972) Detection thresholds for stimuli in humans and monkeys: comparison with threshold events in mechanoreceptive afferent nerve fibers innervating the monkey hand. J Neurophysiol 35(1):122
111. Moy G, Singh U, Tan E, Fearing RS (2000) Human psychophysics for teletaction system design. Haptics-e 1(3):1–20
112. Nakatani M, Sato A, Tachi S, Hayward V (2008) Tactile illusion caused by tangential skin strain and analysis in terms of skin deformation. In: Haptics: perception, devices and scenarios, pp 229–237
113. Nakazawa N, Ikeura R, Inooka H (2000) Characteristics of human fingertips in the shearing direction. Biol Cybern 82(3):207–214
114. Nefs HT, Kappers AML, Koenderink JJ (2002) Frequency discrimination between and within line gratings by dynamic touch. Atten Percept Psychophys 64(6):969–980
115. Nefs HT, Kappers AML, Koenderink JJ (2001) Amplitude and spatial-period discrimination in sinusoidal gratings by dynamic touch. Perception 30:1263–1274
116. Okamura AM, Cutkosky MR, Dennerlein JT (2001) Reality-based models for vibration feedback in virtual environments. IEEE/ASME Trans Mechatron 6(3):245–252
117. Othman MO, Elkholy AH (1990) Surface-roughness measurement using dry friction noise. Exp Mech 30(3):309–312
118. Pai DK, Rizun P (2003) The what: a wireless haptic texture sensor. In: 11th symposium on haptic interfaces for virtual environment and teleoperator systems. HAPTICS 2003, IEEE, New York, pp 3–9
119. Pai DK, Doel K, James DL, Lang J, Lloyd JE, Richmond JL, Yau SH (2001) Scanning physical interaction behavior of 3d objects. In: Proceedings of the 28th annual conference on computer graphics and interactive techniques, pp 87–96
120. Paré M, Elde R, Mazurkiewicz JE, Smith AM, Rice FL (2001) The meissner corpuscle revised: a multiafferented mechanoreceptor with nociceptor immunochemical properties. J Neurosci 21(18):7236
121. Paré M, Behets C, Cornu O (2003) Paucity of presumptive ruffini corpuscles in the index finger pad of humans. J Comp Neurol 456(3):260–266
122. Pasquero J (2006) Survey on communication through touch. Technical report TR-CIM-06.04, Center for Intelligent Machines, McGill University
123. Pastor MA, Day BL, Macaluso E, Friston KJ, Frackowiak RSJ (2004) The functional neuroanatomy of temporal discrimination. J Neurosci 24(10):2585
124. Pasumarty S, Johnson S, Watson S, Adams MJ (2011) Friction of the human finger pad: Influence of moisture, occlusion and velocity. Tribol Lett 44(2):117–137
125. Pataky TC, Latash ML, Zatsiorsky VM (2005) Viscoelastic response of the finger pad to incremental tangential displacements. J Biomech 38:1441–1449
126. Pawluk DTV, Howe RD (1999) Dynamic contact of the human fingerpad against a flat surface. J Biomech Eng 121:605
127. Pawluk DTV, Howe RD (1999) Dynamic lumped element response of the human fingerpad. ASME J Biomech Eng 121:178–184
128. Persson BNJ (2000) Sliding friction: physical principles and applications, vol 1. Springer, Berlin
129. Persson BNJ (2001) Theory of Rubber friction and contact mechanics. J Chem Phys 15(8):3840–3861

130. Phillips JR, Johnson KO (1981) Tactile spatial resolution. II. Neural representation of bars, edges, and gratings in monkey primary afferents. J Neurophysiol 46(6):1192
131. Picard D, Dacremont C, Valentin D, Giboreau A (2003) Perceptual dimensions of tactile textures. Acta Psychol 114(2):165–184
132. Poupyrev I, Rekimoto J, Maruyama S (2007) Mobile apparatus having tactile feedback function. Google patents. US patent 7,205,978
133. Romano JM, Yoshioka T, Kuchenbecker KJ (2010) Automatic filter design for synthesis of haptic textures from recorded acceleration data. In: 2010 IEEE international conference on robotics and automation, IEEE, New York, pp 1815–1821
134. Sathian K, Goodwin AW, John KT, Darian-Smith I (1989) Perceived roughness of a grating: correlation with responses of mechanoreceptive afferents innervating the monkey's fingerpad. J Neurosci 9(4):1273
135. Scheibert J, Leurent S, Prevost A, Debregeas G (2009) The role of fingerprints in the coding of tactile information probed with a biomimetic sensor. Science 323:1503–1506
136. Serina ER, Mote CD, Rempel D (1997) Force response of the fingertip pulp to repeated compression—effects of loading rate, loading angle and anthropometry. J Biomech 30(10):1035–1040
137. Serina ER, Mockensturm E, Mote CD, Rempel D (1998) A structural model of the forced compression of the fingertip pulp. J Biomech 31(7):639–646
138. Shrewsbury M, Johnson RK (1975) The fascia of the distal phalanx. J Bone Jt Surg, Am Vol 57(6):784
139. Siira J, Pai DK (1996) Haptic texturing—a stochastic approach. In: IEEE international conference on robotics and automation, 1996, vol 1, IEEE, New York, pp 557–562
140. Skedung L, Danerlöv K, Olofsson U, Aikala M, Niemi K, Kettle J, Rutland MW (2010) Finger friction measurements on coated and uncoated printing papers. Tribol Lett 37(2):389–399
141. Smith AM, Scott SH (1996) Subjective scaling of smooth surface friction. J Neurophysiol 75(5):1957
142. Smith AM, Chapman CE, Deslandes M, Langlais JS, Thibodeau MP (2002) Role of friction and tangential force variation in the subjective scaling of tactile roughness. Exp Brain Res 144(2):211–223
143. Smith AM, Basile G, Theriault-Groom J, Fortier-Poisson P, Campion G, Hayward V (2010) Roughness of simulated surfaces examined with a haptic tool; effects of spatial period, friction, and resistance amplitude. Exp Brain Res 202(1):33–43
144. Soneda T, Nakano K (2010) Investigation of vibrotactile sensation of human fingerpads by observation of contact zones. Tribol Int 43(1–2):210–217
145. Sreng J, Lécuyer A, Andriot C (2008) Using vibration patterns to provide impact position information in haptic manipulation of virtual objects. In: Haptics: perception, devices and scenarios, pp. 589–598
146. Srinivasan M, Dandekar K (1996) An investigation of the mechanics of tactile sense using two-dimensional models of the primate fingertip. J Biomech Eng 118(1):48
147. Srinivasan MA, Whitehouse JM, LaMotte RH (1990) Tactile detection of slip: surface microgeometry and peripheral neural codes. J Neurophysiol 63(6):1323
148. Srinivasan M, Gulati RJ, Dandekar K (1992) In vivo compressibility of the human fingertip. ASME Adv Bioeng 22:573–576
149. Srinivasan MA (1989) Surface deflection of primate fingertip under line load. J Biomech 22(4):343–349
150. Stevens JC, Harris JR (1962) The scaling of subjective roughness and smoothness. J Exp Psychol 64:489–494
151. Tada M, Pai DK (2008) Finger shell: predicting finger pad deformation under line loading. In: Proceedings of the 2008 symposium on haptic interfaces for virtual environment and teleoperator systems, IEEE Computer Society, Los Alamitos, pp 107–112
152. Tada M, Mochimaru M, Kanade T (2006) How does a fingertip slip?—contact mechanics of a fingertip under tangential loading. In: EuroHaptics 2006, pp 415–420

153. Takasaki M, Kotani H, Nara T, Mizuno T (2005) Transparent surface acoustic wave tactile display. In: Proceedings of the IEEE/RSJ international conference on intelligent robots and systems, pp 1115–1120
154. Talbot WH, Darian-Smith I, Kornhuber HH, Mountcastle VB (1968) The sense of flutter-vibration: comparison of the human capacity with response patterns of mechanoreceptive afferents from the monkey hand. J Neurophysiol 31(2):301
155. Tanaka Y, Horita Y, Sano A, Fujimoto H (2011) Tactile sensing utilizing human tactile perception. In: 2011 IEEE World haptics conference (WHC), IEEE, New York, pp 621–626
156. Tang H, Beebe DJ (1998) A microfabricated electrostatic haptic display for persons with visual impairments. IEEE Trans Rehabil Eng 6(3):241–248
157. Taylor MM, Lederman SJ (1975) Tactile roughness of grooved surfaces: a model and the effect of friction. Atten Percept Psychophys 17(1):23–36
158. Terekhov AV, Hayward V (2011) Minimal adhesion surface area in tangentially loaded digital contacts. J Biomech 44(13):2508–2510
159. Tomlinson SE, Lewis R, Carré MJ (2009) The effect of normal force and roughness on friction in human finger contact. Wear 267(5–8):1311–1318. 17th international conference on wear of materials
160. Tomlinson SE, Carré MJ, Lewis R, Franklin SE (2011) Human finger contact with small, triangular ridged surfaces. Wear 271(9–10):2346–2353. 18th international conference on wear of materials
161. Tomlinson SE, Lewis R, Liu X, Texier C, Carré MJ (2011) Understanding the friction mechanisms between the human finger and flat contacting surfaces in moist conditions. Tribol Lett 41:283–294
162. Tretiakoff O, Tretiakoff A (1977) Electromechanical transducer for relief display panel. Google patents. US patent 4,044,350
163. Van Den Doel K, Pai DK (1998) The sounds of physical shapes. Presence 7(4):382–395
164. Van Den Doel K, Kry PG, Pai DK (2001) FoleyAutomatic: physically-based sound effects for interactive simulation and animation. In: Proceedings of the 28th annual conference on computer graphics and interactive techniques, ACM, New York, pp 537–544
165. Van Doren CL (1989) A model of spatiotemporal tactile sensitivity linking psychophysics to tissue mechanics. J Acoust Soc Am 85(5):2065–2080
166. Van Doren CL (1990) The effects of a surround on vibrotactile thresholds: evidence for spatial and temporal independence in the non-pacinian i (np i) channel. J Acoust Soc Am 87:2655
167. Verrillo RT (1962) Investigation of some parameters of the cutaneous threshold for vibration. J Acoust Soc Am 34:1768–1773
168. Verrillo RT (1963) Effect of contactor area on the vibrotactile threshold. J Acoust Soc Am 35:1962–1966
169. Verrillo RT, Bolanowski Jr SJ (1986) The effects of skin temperature on the psychophysical responses to vibration on glabrous and hairy skin. J Acoust Soc Am 80:528
170. Verrillo RT, Fraioli AJ, Smith RL (1969) Sensation magnitude of vibrotactile stimuli. Atten Percept Psychophys 6(6):366–372
171. Visell Y, Cooperstock JR (2010) Design of a vibrotactile display via a rigid surface. In: 2010 IEEE haptics symposium, pp 133–140
172. Wang Q, Hayward V (2007) In vivo biomechanics of the fingerpad skin under local tangential traction. J Biomech 40(4):851–860
173. Wang Q, Hayward V (2008) Tactile synthesis and perceptual inverse problems seen from the viewpoint of contact mechanics. ACM Trans Appl Percept 5(2):7
174. Warman PH, Ennos AR (2009) Fingerprints are unlikely to increase the friction of primate fingerpads. J Exp Biol 212:2016–2022
175. Watanabe T, Fukui S (1995) A method for controlling tactile sensation of surface roughness using ultrasonic vibration. In: Proceedings of the 1995 IEEE international conference on robotics and automation, vol 1, IEEE, New York, pp 1134–1139

176. Westling G, Johansson RS (1987) Responses in glabrous skin mechanoreceptors during precision grip in humans. Exp Brain Res 66(1):128–140
177. Wettels N, Santos VJ, Johansson RS, Loeb GE (2008) Biomimetic tactile sensor array. Adv Robot 22(8):829–849
178. Winfield L, Glassmire J, Colgate JE, Peshkin M (2007) T-PaD: tactile pattern display through variable friction reduction. In: World haptics 2007, pp 421–426
179. Witney AG, Wing A, Thonnard JL, Smith AM (2004) The cutaneous contribution to adaptive precision grip. Trends Neurosci 27(10):637–643
180. Wu JZ, Welcome DE, Krajnak K, Dong RG (2007) Finite element analysis of the penetrations of shear and normal vibrations into the soft tissues in a fingertip. Med Eng Phys 29(6):718–727
181. Yamamoto A, Nagasawa S, Yamamoto H, Higuchi T (2006) Electrostatic tactile display with thin film slider and its application to tactile telepresentation systems. IEEE Trans Vis Comput Graph 12(2):168–177
182. Yang TH, Pyo D, Kim SY, Cho YJ, Bae YD, Lee YM, Lee JS, Lee EH, Kwon DS (2011) A new subminiature impact actuator for mobile devices. In: 2011 IEEE World haptics conference (WHC), IEEE, New York, pp 95–100
183. Yao HY, Hayward V (2006) An experiment on length perception with a virtual rolling stone. In: Proceedings of Eurohaptics, pp 325–330
184. Yao HY, Hayward V (2010) Design and analysis of a recoil-type vibrotactile transducer. J Acoust Soc Am 128:619
185. Yoshioka T, Bensmaia SJ, Craig JC, Hsiao SS (2007) Texture perception through direct and indirect touch: an analysis of perceptual space for tactile textures in two modes of exploration. Somatosens Motor Res 24(1–2):53–70
186. Yousef H, Boukallel M, Althoefer K (2011) Tactile sensing for dexterous in-hand manipulation in robotics—a review. Sens Actuators A, Phys. doi:10.1016/j.sna.2011.02.038
187. Ziat M, Hayward V, Chapman CE, Ernst MO, Lenay C (2010) Tactile suppression of displacement. Exp Brain Res 206(3):299–310

Chapter 3
Causality Inversion in the Reproduction of Roughness

Abstract When a finger scans a non-smooth surface, a sensation of roughness is experienced. A similar sensation is felt when a finger is in contact with a mobile surface vibrating in the tangential direction. Since an actual finger-surface interaction results in a varying friction force, how can a measured friction force be converted into skin relative displacement? With a bidirectional apparatus that can measure this force and transform it into displacement with unambiguous causality, such mapping could be experimentally established. A pilot study showed that a subjectively equivalent sensation of roughness can be achieved between a fixed real surface and a vibrated mobile surface.

3.1 Opening Remarks

This chapter is based on an article that describes a device able to record and reproduce tactual textures, directly at the fingertip. The instrument comprises a bidirectional piezoelectric transducer that can either be used to record the finger-texture interaction force or, by changing its configuration, be employed to replay it. The high resolution measurements (50 µN) can be acquired over a range of frequencies that covers the tactile perception. Measurements are typically performed when sliding a finger on a surface. The position of the finger is also recorded so that the interaction force can be expressed as a function of space. Textures are reproduced by touching the surface of the device, which in the display mode is mobile. During an exploratory movement, the surface is vibrated, providing the sensation of touching a textured surface, minus the net friction. The same device is used in the following chapters of this book to identify, record, and reproduce several other textures with several aims in mind. A psychophysical calibration method scales the signal amplitude between the measured frictional force due to sliding and the stimulation of the fingertip during reproduction.

Chapter is reprinted with kind permission from Springer Science+Business Media B.V., originally published in [11].

M. Wiertlewski, *Reproduction of Tactual Textures*,
Springer Series on Touch and Haptic Systems, DOI 10.1007/978-1-4471-4841-8_3,
© Springer-Verlag London 2013

3.2 Introduction

Roughness is an important attribute of things we touch [5]. Concomitantly, there is a need for ever increasingly realistic virtual environments that can reproduce the various attributes of objects, including their roughness. To date, the approaches used to simulate roughness include the use of force feedback devices to replicate the microgeometry of surfaces, directly, or by reproducing its effects; see [2] for an extensive survey. Other approaches modulate the friction force that arises when a finger slips on an active surface. To this end, electrostatic fields [12], surface acoustic waves [9], or the squeeze film effect [1, 10], can be employed.

3.2.1 Finger-Surface Interaction

The steady slip of a finger on a surface induces a frictional force. If the surface in question deviates from smoothness, then the interaction force varies over time as a result of a complex interaction taking place between the finger and the surface. Microscopically, the variation is the consequence of the space-and-time-varying traction distribution (i.e. tangential force per unit of contact surface) at the interface between the finger and the surface. The traction distribution depends on the relative geometries of these two bodies, on the materials they are made of, and on the possible presence of fluids and foreign bodies.

In spite of this complexity, integration of traction over the (unknown) contact surface results in a net force that can be measured. It is known that the variations of this force correlate strongly with a sensation of roughness [7, 8]. While there is much debate regarding the manner in which the nervous system mediates the sensation of roughness peripherally and centrally, there is evidence that a variety of mechanisms are at play. Because of the multiplicity of these mechanisms, diverse stimulation methods can contribute to elicit roughness, see [6] among others.

3.2.2 Present Study

The present paper explores the possibility of stimulating the cutaneous system in order to create roughness sensations through the simplest method possible: that of vibrating tangentially a smooth surface in non-slipping contact with the finger, as the finger undergoes net motion. Yet, when it comes to design a display based on this idea, this simple approach poses a basic question which must be clearly answered: During tactile exploration, does the finger-surface interaction force "cause" the finger to deform or does the deformation "cause" the interaction force? Visual or auditory displays, by-and-large, radiate the same energy regardless of how they are looked at or listened to, so the causality is clear, but for haptic displays the causality question cannot be answered so easily, see [4] for elementary notions. The same question can be rephrased as follows: Should the measurement and the simulation be based on the skin displacement or on the force applied to it?

3.2.3 Bidirectional Apparatus

To study this question, we build an apparatus that unambiguously establishes a causal relationship between the measurement and the stimulation by operating both as a sensor and as an actuator. In these two cases, the device was engineered to be very stiff, that is, five orders of magnitude stiffer than a fingertip. This way, when used as a sensor, the interaction force is known regardless of the finger movements and deformations; when used as an actuator, displacement is specified independently from the interaction force. To complete the symmetry, during recording operations, the sensor is fixed with respect to the ground and the finger slips on a rough surface. During restitution, the actuator is mounted on a slider and remains fixed relatively to the scanning finger touching a flat surface. In both modes, the device operates with a bandwidth spanning from 20 to 600 Hz, and has a maximum displacement of 0.2 mm in actuator mode, thereby covering the range useful for conveying roughness.

3.2.4 Main Result

We performed a preliminary psychophysical experiment aimed at finding the subjective equivalence of roughness elicited by a rapidly varying measured force or by an imposed displacement, hence realizing a causality inversion between the measurement and the display. This approach is in contrast with the one employed with conventional haptic devices where a force is measured, or computed, and then specified with impedance devices; or where a displacement is computed and then specified with admittance devices. It was found that, indeed, such subjective equivalence of roughness could be established.

3.3 Apparatus

Referring to Figs. 3.1a and 3.3a, the apparatus comprises a rigid plate, A supported at one end by a low stiffness blade, B, and connected to a multilayer piezoelectric circular bender (CMBR07, Noliac Group A/S, Kvistgaard, Denmark), C, at the other. As a sensor, a textured surface is glued to the plate and during scanning the interaction force is measured within a very large dynamic range. As an actuator, the assembly is mounted on a linear guide and the smooth plate is vibrated tangentially.

3.3.1 Sensor Operation

The piezoelectric bender converts tangential forces due to the interaction with the finger into electric charges. These charges are transformed into voltage by an instrumentation amplifier (LT1789, Linear Technology Corp., Milpitas, CA, USA) as

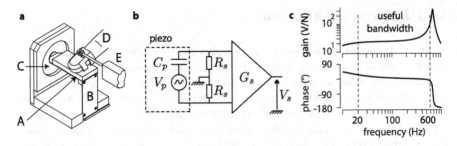

Fig. 3.1 Sensor operation. **a** Setup. The finger position, measured by E, and the interaction force, measured by C are recorded when the finger, D, slips on the surface, A. **b** Signal conditioning. **c** Frequency response of the sensor

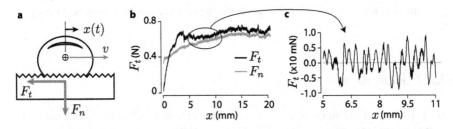

Fig. 3.2 Measurement. **a** The components F_n and F_t of the interaction force during scanning. $x(t)$ is measured by a LVDT. **b** Typical force measurement from the conventional force sensor. **c** Wide range dynamic measurement of sliding interaction measured by the piezoelectric sensor

shown in Fig. 3.1b. The signal is then digitized by a 16-bit data acquisition board (PCI-6229, National Instruments Corp., Austin, TX, USA). The piezoelectric transducer acts like a generator V_p in series with a capacitor C_p and a charge resistance R_s. The RC circuit corresponds to a 20 Hz high-pass filter. Such charge-based force sensor is capable of a very high dynamic range response unachievable with conventional strain-gauge-based force sensors.

The position of the finger, D, is measured with a linear variable transformer transducer (LVDT), E, (SX 12N060, Sensorex SA, Saint-Julien-en-Genevois, France) fastened to the fingernail. The response, Fig. 3.1c, shows a sensitivity of 26 V/N in the range from 20 Hz to 600 Hz. The range is limited upward by the mechanical natural resonance. Output noise is lower than 20 μN/√Hz and 16-bit digital conversion provides 50 μN of resolution at a 2 kHz sampling rate. The high stiffness of the bender (70×10^3 N/m) ensures that the small deformation hypothesis is valid. Low frequency force components are measured by a conventional force sensor (Nano 17, ATI Industrial Automation, Inc., Apex, NC, USA) mounted on the load path between the assembly and a firm mechanical ground.

The interaction force components F_t and F_n and finger position $x(t)$ are acquired by the sensor during scanning, see Fig. 3.2a. A typical measurement is seen in Fig. 3.2b. Notice how the tangential force F_t rises at the beginning of the motion

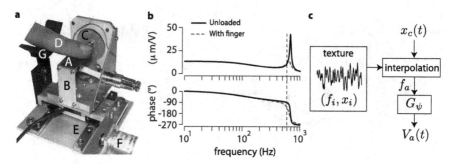

Fig. 3.3 Actuator operation. **a** The stimulator, C, mounted on a linear stage, E, is fixed relatively to the finger, D. **b** Frequency response of the none loaded actuator (*black line*) and with a finger pushing at 1 N (*dash line*). **c** Control

and then oscillate around a value. The high-pass filter preserves the variation of the tangential force occurring within a wide dynamic range that a conventional force sensor would be unable to resolve, as shown in Fig. 3.2c. The initial stick to slip transition and ensuing transients have been edited out for clarity. This diagram is representative of the rich variations of the friction force due to a finger slipping on a periodic grooved surface.

3.3.2 Actuator Operation

Referring to Fig. 3.3a, when used as an actuator, the assembly is disconnected from the grounded force sensor and placed on a slider, E. Its position is measured with a 7.5 μm linear resolution using an incremental encoder, F (Model 2400, Fritz Kübler GmbH, Villingen-Schwenningen, Germany). The participant's third phalanx rests on a cradle, G, connected to the slider so that the fingertip rests on the active surface, A. As the finger scans to and fro, the transducer is driven by a voltage amplifier (Apex Precision Power PA86U, Cirrus Logic Inc., Austin, TX, USA) such that the skin in contact with the active surface is entrained by its oscillations without slip.

In order to ascertain performance, the output displacement was measured with a laser telemeter (LT2100, Keyence Corp., Osaka, Japan). The response, Fig. 3.3b, shows that the system is able to produce a displacement of ± 20 μm/V from DC to 600 Hz, limited by the system's natural resonance. For a 5 V input, the actuator is able to achieve a maximum displacement of 100 μm.

The actuator is driven by a 2 kHz periodic realtime thread that reads the encoder position $x_c(t)$, interpolates a force value f_a from a given texture profile, multiplies it by a gain G_ψ and refreshes the amplifier output $V_a(t)$, see Fig. 3.3c. This control thread runs under the LabVIEW™ environment on an ordinary computer equipped with the digital input-output board already mentioned.

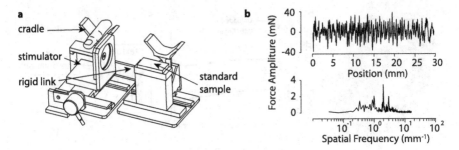

Fig. 3.4 **a** Experimental setup. **b** Stimulus and its spectrum. The scanning process transforms a simple surface waveform into a complex, broadband force signal

3.4 Experimental Procedure for Perceptual Calibration

Having unambiguously converted a varying interaction force into a skin displacement during the scanning of a surface by means of an apparatus designed to establish robust causal relationships not achievable with conventional haptic devices, the question now arises of the value of the conversion factor that could elicit an equivalent sensation of roughness. Furthermore, if such a factor exists, does it vary from person to person? To address these questions, a calibration procedure was carried out with six participants in order to establish the point of subjective equivalent roughness between the natural texture and its simulated version. A 2-alternative forced choice, constant stimuli procedure was employed to find the gains $\hat{G}_\psi$ that would elicit an equivalent sensation of roughness.

3.4.1 Stimuli

The standard stimulus was a triangular grooved grating of 1 mm spatial period with 0.1 mm of depth. Without relative motion, the roughness of this texture was not perceptible. The scanning force with this grating was measured using the sensor described earlier with the help of a "standard" participant. During recording, the speed v and the normal force F_n were held constant with a 10 % tolerance. The signal was processed as described in Sect. 3.3.1, then normalized to ± 0.5 V. The filtered signal, expressed in Newton, and its Fourier transform are shown in Fig. 3.4b. The comparison stimulus was provided by the stimulator described in Sect. 3.3.2. Figure 3.4a illustrates the precautions that were taken so that both stimuli were presented in exactly the same conditions: (a) The participants had their proximal phalanx resting in cradles connected to sliding guides so that the fingertip rested on the grooved texture or stimulator in same manner; (b) Both surfaces were made in polycarbonate; (c) The two sliders were mechanically connected so the same inertia and the same friction was felt for the standard and the comparison stimuli. The experimental setup was hidden by a curtain to avoid visual bias. Subjects wore sound isolation headphones (model K518, AKG Acoustics, Harman International Industries) playing white noise.

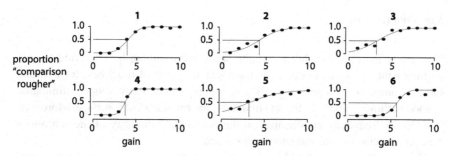

Fig. 3.5 Results of each participants and their sigmoid fitting

Fig. 3.6 Value of the Point
of Subjective Equivalence of
each subject

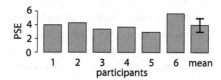

3.4.2 Subjects and Procedure

Six volunteers participated in the experiment, two female and four male, all right-handed, aged from 24 to 31 years. They were from CEA LIST and two of them were familiar with haptic technology. Their hand was guided to explore the setup and after short instructions they were asked to judge whether the standard or the comparison stimulus was rougher and to give an answer via a keystroke. Like in [3], no definition of roughness was given except that "roughness is the opposite of smoothness". Neither training nor feedback was provided during the tests. Gain G_ψ was randomly chosen in a range of 1 to 10. Each value was tested at least 10 times.

3.5 Results

Subjects responded to gain changes following a typical psychometric curve, as shown in Fig. 3.5. The data were fitted with a cumulative Gaussian distribution $f(x) = 0.5[1 + \text{erf}((x - \mu)/\sqrt{2}\sigma)]$ where x is the gain, μ the mean gain and σ^2 the variance. The data fitting was achieved using a nonlinear least-square fitting procedure.

The point of subjective equivalence (PSE) was extracted from the gain that corresponds to a 50 % probability of judging the comparison rougher than the standard. Figure 3.6 shows the distribution of PSE's. The average across subjects is $\bar{G}_\psi = 3.99$ with a standard deviation of 0.93.

3.6 Discussion

The results indicate that the interaction force variation can be converted to skin displacement variations to elicit an equivalent sensation of roughness for a virtual surface compared to a real one. As a result, this particularly simple stimulation method is shown to be effective at simulating the roughness of a surface. Moreover, participants frequently commented on the perceptual similarity of the sensations themselves between real and simulated surfaces.

These results support the idea that as far as fine textures are concerned, spatial information can be completely eliminated from the simulation, yet, the conscious experience can be that of a non-smooth surface. While similar observations have frequently be reported in the past, our experiments, given to the care that we put in controlling the causality as well as the quality of the transmitted signals, make it now possible to quantify the conditions under which such phenomenon occur. Another aspect of our results worthy of some comments is the relative constancy of the conversion factor among individuals. Of the six individuals who lent themselves to the experiment, five obtained very similar numbers. Only one required a significantly higher displacement stimulus to achieve an equivalent level of roughness. Our efforts will be directed in the future at understanding these individual differences.

3.7 Conclusion

With the help of a carefully engineered sensor, sliding frictional forces could be acquired within a very high dynamic range. The same device was turned in a stimulator having, by construction, a compatible dynamic range that could convert this frictional force into a displacement able to provide a simulated sensation.

This study has so-far considered only one texture for perceptual calibration. We plan to investigate other aspects of texture signals, such as spectral content, in addition to amplitude, and to study the conditions under which perceptual equivalence can be achieved. The distribution of roughness perception across gain values was found to be a monotonic function. As a result, one could employ fast calibration procedures such as accelerated staircase methods as in [3].

A final implication of the present experiment is the possibility to replace force feedback stimulation by cutaneous displacement stimulators which may lend themselves to more favorable engineerings tradeoffs, particularly with subminiature devices. Such miniature devices could for instance be embedded in the gripping surfaces of conventional force feedback devices.

Acknowledgements   The author would like to thank Margarita Anastassova for her helpful comments on the experimental setup. This work was supported by the French research agency through the REACTIVE project (ANR-07-TECSAN-020).

References

1. Biet M, Giraud F, Lemaire-Semail B (2007) Squeeze film effect for the design of an ultrasonic tactile plate. IEEE Trans Ultrason Ferroelectr Freq Control 54(12):2678–2688
2. Campion G, Hayward V (2008) On the synthesis of haptic textures. IEEE Trans Robot 24(3):527–536
3. Campion G, Hayward V (2009) Fast calibration of haptic texture synthesis algorithms. IEEE Trans Haptics 2(2):85–93
4. Hogan N (1985) Impedance control: an approach to manipulation. J Dyn Syst Meas Control 107:1–7
5. Lederman SJ, Taylor MM (1972) Fingertip force, surface geometry, and the perception of roughness by active touch. Percept Psychophys 12(5):401–408
6. Maeno T, Otokawa K, Konyo M (2006) Tactile display of surface texture by use of amplitude modulation of ultrasonic vibration. In: Proceedings of the IEEE ultrasonics symposium, pp 62–65
7. Smith AM, Chapman CE, Deslandes M, Langlais JS, Thibodeau MP (2002) Role of friction and tangential force variation in the subjective scaling of tactile roughness. Exp Brain Res 144(2):211–223
8. Smith AM, Basile G, Theriault-Groom J, Fortier-Poisson P, Campion G, Hayward V (2010) Roughness of simulated surfaces examined with a haptic tool; effects of spatial period, friction, and resistance amplitude. Exp Brain Res 202(1):33–43
9. Takasaki M, Kotani H, Nara T, Mizuno T (2005) Transparent surface acoustic wave tactile display. In: Proceedings of the IEEE/RSJ international conference on intelligent robots and systems, pp 1115–1120
10. Winfield L, Glassmire J, Colgate JE, Peshkin M (2007) T-PaD: tactile pattern display through variable friction reduction. In: World haptics 2007, pp 421–426
11. Wiertlewski M, Lozada J, Pissaloux E, Hayward V (2010) Causality inversion in the reproduction of roughness. In: Kappers AML, van Erp JBF, Bergmann Tiest WM, van der Helm FCT (eds), Haptics: generating and perceiving tangible sensations—international conference, EuroHaptics 2010, Amsterdam, July 8–10, 2010, pp. 17–24
12. Yamamoto A, Nagasawa S, Yamamoto H, Higuchi T (2006) Electrostatic tactile display with thin film slider and its application to tactile telepresentation systems. IEEE Trans Vis Comput Graph 12(2):168–177

Chapter 4
The Spatial Spectrum of Tangential Skin Displacement

Abstract The tactual scanning of five naturalistic textures was recorded with an apparatus capable of measuring the tangential interaction force with a high degree of temporal and spatial resolution. The resulting signal showed that the transformation from the geometry of a surface to the force of traction, and hence to the skin deformation experienced by a finger is a highly nonlinear process. Participants were asked to identify simulated textures reproduced by stimulating their fingers with rapid, imposed lateral skin displacements as a function of net position. They performed the identification task with a high degree of success, yet not perfectly. The fact that the experimental conditions eliminated many aspects of the interaction, including low-frequency finger deformation, distributed information, as well as normal skin movements, shows that the nervous system is able to rely on only two cues: amplitude and spectral information. The examination of the "spatial spectrograms" of the imposed lateral skin displacement revealed that texture could be represented spatially despite being sensed through time and that these spectrograms were distinctively organized into what could be called "spatial formants". This finding led us to speculate that the mechanical properties of the finger enables spatial information to be used for perceptual purposes in humans without any distributed sensing, a principle that could be applied to robots.

4.1 Opening Remarks

The previous chapter described a device for reproducing roughness and texture sensations on the fingertip as well as an experiment that validates the approach. The present chapter provides a more detailed description of the apparatus, the development of a electromechanical model, and a calibration procedure to assess the quality of the measurements and of the stimulation. Next, a psychophysical experiment was carried out to test the ability of participants to matched the reproduction of five virtual textures with their real counterpart. A second experiment was aimed at testing the ability of subject to discriminate the spatial frequencies of virtual gratings. The results were compared with published results found with actual samples. This chap-

---

Chapter is reprinted with kind permission from IEEE, originally published in [43].

ter also hypothesizes the existence of a spatial code for texture roughness which is not referenced to the skin but to the touched surface.

4.2 Introduction

Texture—the organized deviation from smoothness of the surface of objects—typically is first apprehended visually but once contact is made with the hand, touch must take charge. Katz, in 1925, noted that there are two ways in which organisms can become tactually aware of the texture of objects [14]. One way is to determine directly the relevant spatial features of the geometry of a touched surface. To illustrate how this could be done, consider a deeply grooved grating, such as a knurled knob. Under reasonable normal static loading, the skin interacts with such a surface through a collection of minute contact surfaces. Assuming that the sensory apparatus is able to detect these individual contact surfaces, then presumably a coarse notion of the surface geometry can be acquired. If such surface has any degree of fineness, however, the individual contacts become so numerous that such strategy becomes highly implausible. Most psychologists and neurophysiologists agree with Katz that the experience of surface texture must result from mechanical signals brought about by finger sliding that change through time, in addition to mechanical signals that vary through space.

To the haptics engineer interested in devices and transducers able reproduce tactile and haptic sensations, these observations are very significant since the overriding objective is to extract from the complexity of the ambient physics those aspects that are the most significant to the perceiver and to discard the others in the name of technical feasibility.

Of course, tactual texture is an ill-defined notion. In a single sentence, it is hard to discuss the sensations caused by rough burlap, those resulting from finely machined bronze, or those derived from the velvety skin of an apricot. To make things worse, from a physical view point, and even restricting attention to hard materials, texture and roughness can be characterized in many different ways that also depend on the method used to measure it [2, 36]. With soft materials the situation is even more inextricable.

The many studies in the psychophysics of texture and roughness perception unfortunately contribute little insight to the haptics engineer because these studies rarely speak of the same quantities, although there is a general agreement that roughness has perceptual significance [6, 8, 9, 12, 13, 17, 18, 23, 29, 31, 34], even if it is nearly impossible to define it unambiguously from the physical characteristics of the touched object [4].

If roughness, and more generally, if tactual texture is hard to discuss directly from the physics of an object, then perhaps a more productive approach from the view point of interface design would be to focus on the characteristics of the mechanical interaction of the skin with an object, although the prospects for identifying simple signals are rather bleak at first sight. The finger is a soft, highly deformable object which, besides its complex detailed geometry, exhibits several types of nonlinearities that are manifest at different length scales of interaction with surfaces [1, 26, 27, 39, 40]. Even under the extremely simplified assumption of linear visco-elasticity and perfectly clean contacts free of foreign bodies and liquids, the contact of deformable bodies with rough surfaces gives rise to theories of considerable complexity that are unlikely to yield simple interaction models [28]. These observations justify the measurement-reproduction approach adopted in this study.

4.3   Roughness and Texture in Manufacturing and Virtual Reality

Numerous industrial processes, from mirrors to roads, depend on the measurement of roughness. It is achieved using profilometers based on slow mechanical scanning with a sharp stylus (the tip radius can be as small as a few nanometers) or by optical methods (confocal microscopy, laser triangulation, interferometry). Reporting roughness is mostly a function of the intended application. In part machining, for instance, roughness is traditionally characterized in terms of the relative heights of a set of asperities specifying their standardized moments: 0th, 1st, 2nd, 3rd. Interestingly, the latter measures report zero roughness for any regular grating and therefore cannot be applied to perceptual studies. Other measures report the statistics of the peak-to-valley distances of sets of asperities which makes them more relevant. Some measures consider autocorrelation, some account for spatial wavelength or for extrema density. Some others take into consideration the magnitude of the slopes of asperities, and yet others their curvature. The later measure is probably one of the most relevant to tactual roughness of these different approaches.

The measurement process is typically slow (minutes, hours) and provides details that are not necessarily relevant to tactual sensing. On the other hand, it is an everyday experience that the roughness of a surface can be felt, or that two textured surfaces can be discriminated, or even that a wood grain can be identified in a fraction of a second by the scanning finger.

These observations have let researchers in virtual reality to adopt the more expeditious method used by humans to sense texture, rather than to rely on industrial-type methods. Examples of this approach can be found in [25] where the scanning interaction force is measured, in [46] where the scanning acceleration of a stylus is measured, or in [38] where the scanning velocity is measured. The reader is referred to a recent survey where 50 articles on the subject are commented [16].

For texture reproduction, the most widely adopted approach is the force-feedback device with a position-dependent textured virtual wall, also extensively surveyed in [16]. A more recently introduced technique is to modulate the friction force between the finger and a mechanically grounded active surface. The friction force modulation can be achieved, for instance, by electrostatic fields [45], ultrasonic amplitude modulation [20], surface acoustic waves [35] or with the squeezed film effect [5, 44].

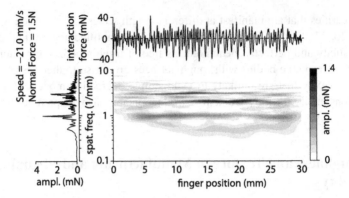

Fig. 4.1 Spatial spectrogram produced by sliding a finger on a perfectly periodic triangular grating. The methods used to construct such plots are described in detail later in this article. For now, it can be appreciated that the transformation from a triangular profile to a force signal is far from straightforward. A triangular wave has only odd harmonics. While the fundamental "formant", or spectral peak, at 1 mm^{-1} is present, it is actually weaker than the first even-harmonic spectral peak. Notice also the present of energy in the sub-harmonic frequencies. These are the hallmarks of a nonlinear transformation

4.4 Design Motivation for a Texture Transducer

The above considerations led us to engineer a new device capable of accurate measurement and reproduction of a surface-finger interaction, having a bandwidth and a dynamic range able to do justice to the biomechanics and sensory performance of the human finger. This device is already briefly described in reference [41] where it was shown that it was able to provide perceptually equivalent sensations of roughness between a virtual and a real surface. The surface used in these preliminary experiments was a "simple" triangular grating of spatial period 1.0 mm with groove depth 0.1 mm. Although the surface was periodic, the force of interaction during sliding turned out to be a complex, broadband signal having a complicated harmonic signature which can be appreciated by consulting Fig. 4.1 and caption. The transformation from geometry to signal is highly nonlinear, a fact that is hardly surprising considering that friction is the primary phenomenon involved [42].

In the present article, we describe this device in greater detail and we employ it in a experiment where it is used to reproduce various textures. We show that the textural recording-reproduction obtained with this device is of sufficient quality to enable several participants to correctly match a virtual surface with a real surface included in a set of five. The mechanical consequences of net friction were eliminated by the transduction process and so was distributed skin deformation within the finger contact area. As a result, the apparatus reproduced accurately the oscillatory components of the skin tangential displacement at the exclusion of other mechanical consequences of sliding a finger over a rough surface.

A particular feature of our device is that the same mechanical structure was used in the sensor and actuator modes. It is based on the piezoelectric effect which, as is well known, is reversible. In sensor mode it operates like a high-quality, stiff force

Fig. 4.2 **a** Transducer schematic. **b** Cross section of the system at rest (*solid lines*) and during deformation (*dashed lines*)

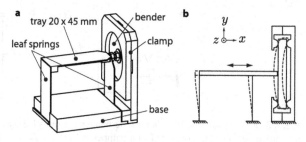

sensor. In actuator mode it provides accurate isometric stimulation to the skin. The questions regarding the reciprocal signal causalities are discussed in [41].

4.4.1 Performance Considerations

The device should be orders of magnitude stiffer than the fingertip to provide unambiguous measurement and stimulation, noting that the converse possibility is considerably harder to achieve due to the difficulties met in reducing the effects of inertia to sub-threshold levels [22].

Other design considerations include the level at which interaction forces should be resolved. In absence of knowledge on the smallest dynamic forces able to stimulate the skin, an estimate can be obtained by considering that the elasticity of the fingertip is roughly of the order of 10^3 N m^{-1} and that a detectable skin displacement is of the order of 10^{-7} m. One could infer that the sensor should resolve 10^{-4} N, which is far beyond the reach of commercial strain-gauge force sensors. In terms of actuator displacement, similar considerations indicate that 10^{-4} m would be needed to create the 10^{-1} N peak-to-peak force oscillations that can be encountered when stroking texture as can be seen from Fig. 4.1. This requirement has been, in hindsight, the hardest to meet and, due to saturation, has somewhat limited the scope of our investigations. Finally, it is commonly accepted that a 500 Hz bandwidth is needed to represent tactile interactions. Interestingly, this figure was actually proposed by Katz almost a century ago [14].

4.4.2 Description

The main components, shown in Fig. 4.2a, comprise a multilayer, 40 mm piezoelectric disk-bender (CMBR07, Noliac Group A/S, Kvistgaard, Denmark) connected to a 20 mm-wide tray that can hold a sample. The bender is clamped vertically by two epoxy ridges of semi-circular section that apply uniform pressure on the bender. A treaded rod connected to the hollow center of the bender transmits motion to the tray which is linearly guided by a flexure made of two leaf springs. Connection

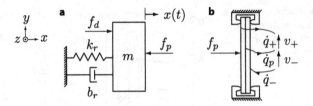

Fig. 4.3 a The plate, the sample and the bender are modeled by a mass m. It is suspended by a spring of stiffness k_r connected to a damper b_r. Forces f_p and f_d represent the piezoelectric actuator and finger interaction forces, respectively. **b** The piezoelectric effect causes charges to appear on the electrodes as a result of displacement

to the bender is realized by two Delrin© washers that can tolerate $\pm 0.5°$ of misalignment. The texture samples are bonded to the tray using double-sided adhesive tape.

During sensor operation, the interaction forces induce flexural deformations of the blade along x and in the piezoelectric element, as indicated in Fig. 4.2b. Through the piezoelectric effect, the deformation of the ceramic causes charges to appear on the electrodes that are picked up by an instrumentation amplifier. Conversely, when applying a voltage to the electrodes of the bender, the piezoelectric effect causes the transducer to operate as an actuator. In this case, displacements of the tray impose deformations in the fingertip resting on it.

4.4.3 Transducer Modeling

Since the mechanical constitution of the transducer is common to the sensor and to the stimulator, their models include the same lumped parameters. They differ only by the electronics. In sensor mode, a high gain, low noise instrumentation amplifier collects charges and convert it into readable voltage, whereas in actuator mode, a high voltage amplifier, with a low output impedance is used to drives voltage on the electrodes of the piezoelectric bender and therefore the tray's displacement.

4.4.3.1 Mechanical Flexure

The flexure acts like a mass-spring-damper system with stiffness k_r according to Fig. 4.3a. Damping due to internal and external friction is by and large dominated by losses in the bender. It arises mostly from hysteresis in the piezoelectric material. As further discussed later, it is reasonable to represent it by viscous damping. The inertial term corresponds to the equivalent moving masses of the tray and of the bender. The actuator force is shown as an external force, f_p, acting in the opposite direction of x. Another external force, f_d, models the finger interaction through its contact with the sample.

Applying Newton's second law and converting to the Laplace domain gives

$$\left(ms^2 + b_r s + k_r\right) X(s) = -F_p(s) + F_d(s), \tag{4.1}$$

where $X(s)$, $F_p(s)$ and $F_d(s)$ represent the Laplace transform of the variables $x(t)$, $f_p(t)$ and $f_d(t)$.

4.4.3.2 Static Constitutive Relationships

The Y-poled bimorph piezoelectric element has two external electrodes plus one central electrode located in the neutral fiber. Bending deformation results in the compression of one layer and traction of the other. Layers are polarized which creates charges q_+, q_p and q_- through the piezoelectric effect. Operating as an actuator, imposed voltages v_+ and v_- push the charges on the armatures to induce axial deformation as a result of the radial strain. The linear, static constitutive relationships can be expressed in matrix form as follows [33],

$$\begin{pmatrix} x \\ q_+ \\ q_- \end{pmatrix} = \begin{pmatrix} 1/k_p & \beta & \beta \\ \beta & C_p & 0 \\ \beta & 0 & C_p \end{pmatrix} \begin{pmatrix} f_p \\ v_+ \\ v_- \end{pmatrix}, \tag{4.2}$$

where $\beta = \delta_{\max}/2v_{\max}$ is the ratio of the largest unloaded deflection to the total maximum operating voltage applied to one layer, k_p is the flexural stiffness in open circuit and C_p is the capacitance of one piezoelectric layer when no stress is applied. In this model, the dynamic parameters like mass and damping are not taken into account.

4.4.3.3 Transfer Function in Sensor Mode

Only one layer is used. The voltage generated by an external force can be written from (4.2) by summing the piezoelectric induced voltage, v_p, with the voltage due to the circulation of charges,

$$v_+ = \frac{q_0 + \int \dot{q}_+ \, dt}{C_p} - \frac{\beta}{C_p} f_p = \frac{q_+}{C_p} + v_p.$$

The transducer acts electrically like a voltage generator in series with a capacitor C_p. The generated voltage, v_+, is amplified by a high input impedance ($10^{12}\ \Omega$) instrumentation amplifier (LT1789, Linear Technology Corp., Milpitas, CA, USA), see Fig. 4.4. Load resistances, R_s, combined with the capacitor form a first-order high-pass filter which can be expressed in the Laplace domain by

$$V_+ = -\frac{2\beta R_s s}{1 + 2R_s C_p s} F_p = \frac{2R_s C_p s}{1 + 2R_s C_p s} V_p. \tag{4.3}$$

Fig. 4.4 Schematic of the sensor circuit. The electrode of the upper layer is connected to an instrumentation amplifier. Resistors create a high-pass filter

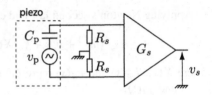

Fig. 4.5 Circuit in actuator mode. Upper and lower electrodes are connected to fixed voltages $\pm v_{\max}$. The power amplifier drives the central electrode voltage, v_p

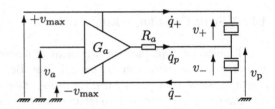

Neglecting the contribution of v_p to the mechanical behavior, the mechanical constitutive equation is $x(t) = 1/k_p\, f_p(t)$. Combining it with (4.3), $V_s(s) = G_s V_+(s)$ gives the output voltage, $V_s(s)$, as a function of the displacement $X(s)$:

$$V_s(s) = -G_s \beta k_p \frac{2R_s s}{1 + 2R_s C_p s} X(s).$$

Using (4.1), the transfer function of the sensor, $H_s(s)$, becomes

$$H_s(s) = \frac{V_s(s)}{F_d(s)} = \frac{-2G_s R_s k_p \beta s}{(1 + 2R_s C_p s)(ms^2 + b_r s + k_r + k_p)}.$$

4.4.3.4 Transfer Function in Actuator Mode

The bender is connected to a power source (PA86U, Cirrus Logic Inc., Austin, TX, USA) which drives the central electrode as in Fig. 4.5. The amplifier is connected voltage-mode with a gain $G_a = 20$. A resistor, R_a, in series with the output tunes the frequency response since a low pass filter is formed with the capacitance of the piezoelectric element.

By application of Kirchhoff's law at the output node,

$$\dot{q}_+ + \dot{q}_p = \dot{q}_-, \tag{4.4}$$

with $\dot{q}_+ = C_p \dot{v}_+$, $\dot{q}_p = (1/R_a)(G_a v_a - v_p)$ and $\dot{q}_- = C_p \dot{v}_-$. Using these values in (4.4) and substituting $v_+ = v_{\max} - v_p$ and $v_- = v_p + v_{\max}$ yields in the Laplace domain,

$$\frac{1}{R_a}(G_a V_a - V_p) = C_p(V_p + V_{\max})s - C_p(V_{\max} - V_p)s,$$

finally giving,

$$V_p = \frac{G_a}{1 + 2R_a C_p s} V_a. \tag{4.5}$$

The power stage acts as an amplifier of gain G_a with a first-order low-pass filter of cutoff frequency $\nu_{cut} = 1/(4\pi R_a C_p)$. The first line of (4.2) combined with (4.1) gives

$$\left(ms^2 + b_r s + k_r + k_p\right)X(s) = 2\beta k_p V_p(s) + F_d(s). \tag{4.6}$$

The two last lines of (4.2) can be simplified since the voltage driver supplies and draws charges as necessary such that (4.2) becomes

$$q_+ = C_f V_+ \quad \text{and} \quad q_- = C_f V_-.$$

The transfer function of the unloaded stimulator is found by combining (4.5) with (4.6),

$$H_a(s) = \frac{X(s)}{V_a(s)} = \frac{G_a k_p \beta}{(1 + 2R_a C_p s)(ms^2 + b_r s + k_r + k_p)}.$$

4.4.4 Identification

The model parameters could be initially estimated from the data provided by manufacturers, as well as from the design of the leafs and of the tray. Parameter identification was then performed to obtain a better model and to take into account nonlinearities and parameter deviations from their manufacturing specification. Because the mechanical parameters are common to the sensor and the actuator, it is more convenient to identify the system first in actuator mode.

4.4.4.1 Actuator Mode

The frequency response was determined using a frequency sweep from 10 Hz to 1000 Hz, applying 4 Vpp voltage signal, v_a (using a digital-to-analog converter PCI-6229, National Instruments Corp., Austin, TX, USA). Output displacement was measured using a laser telemeter (LT2100 with LC2210, Keyence Corp., Osaka, Japan). At each frequency, amplitude and phase were measured after a 200 ms pause to let the transients subside. The response was determined under the following conditions: unloaded, with a finger resting on the tray (normal force ≈ 0.5 N) and with a finger pushing down the plate (normal force ≈ 1 N). The result is shown in Fig. 4.6. The system exhibits the intended natural resonance at 500 Hz followed by a small un-modeled resonance at 800 Hz.

Since the actuator is two order of magnitude stiffer than the finger, finger loading has a negligible impact on response in a 10–400 Hz band. At resonance, however,

Fig. 4.6 Frequency response of the actuator. Measurement are in *grey dot* and the model in plain *black*

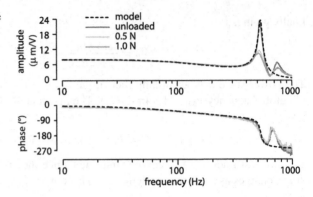

Table 4.1 Electro-mechanical parameters

Mechanical	Electrical
$\delta = 82.35 \ \mu m$	$C_p = 818 \ nF$
$k_p = 76.47 \times 10^3 \ N\,m^{-1}$	$v_{max} = 100 \ V$
$m = 6.8 \ g$	$G_s = 100$
$b_r = 4.49 \ N\,mm^{-1}\,s^{-1}$	$R_s = 12 \ k\Omega$
$k_r = 4.05 \times 10^{-3} \ N\,m^{-1}$	$G_a = 20$
	$R_a = 680 \ \Omega$

damping due the fingertip causes a 3 dB attenuation of the resonant peak. In the experiments, caution was taken to roll-off the signal with 3 dB attenuation at 500 Hz, flattening the response. The effects of the finger damping as well as of the second resonance can therefore be neglected. Least-square fitting ($R^2 = 0.85$) provided the parameters shown in Table 4.1. As can be seen from the figure, the model and the uncorrected system responses are graphically indistinguishable up to 500 Hz.

It is known that piezoelectric ceramic transducers have significant hysteresis which affects the quasi-static and the dynamic responses. Figure 4.7 plots the response of the transducer to 0.1 Hz sinusoidal 10 V peak-to-peak amplitude signal showing a 16 % hysteresis. The hysteresis introduced by piezoelectric ceramics is of non-saturating type and hence introduces small amplitude distortion of no consequence in our experiments, since the minor loops are very small. It does introduce constant phase lag of 8° which, at a given frequency, can be represented as linear damping [10]. It is the actuator hysteresis that accounts for the nicely damped resonance of the system at 500 Hz but is neglected in the low frequencies. In summary, the actuator is capable of a maximum peak-to-peak displacement of 200 μm, with a quasi-static gain of 20 μm/V in the range from DC to 500 Hz.

4.4.4.2 Sensor Mode

A known external force, calibrated using a conventional force sensor (Nano 17, ATI Industrial Automation, Apex, NC, USA), was applied to the sensor. This force was

Fig. 4.7 Quasi-static measurement of the actuator response (*gray circles*). Data show a non saturating hysteresis that can be approximate by a 8° ideal phaser (*black line*)

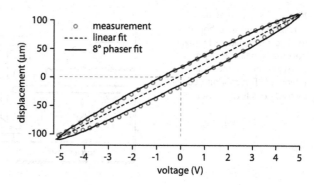

Fig. 4.8 Fit of the sensor model with actual measurements

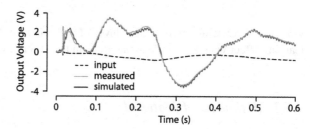

used as an input for the model described earlier. Figure 4.8 the fit of the model with the measurement ($R^2 = 0.91$). The simulated output is 10 times more noisy than the actual measurement from the sensor because of the noisy input measurements from the strain-gauge force sensor.

The model predicts a sensitivity of 26 V N^{-1} for a gain of $G_s = 100$ in the bandwidth 10–500 Hz with the resistor R_s set to 12 kΩ. Like in the actuator mode, the sensor has a natural resonance at 500 Hz. The signal is acquired with the data acquisition board already mentioned. With 16 bits of resolution, the force signal can be measured with 10^{-5} N resolution. The experimentally measured noise floor is as low as 25 μN ($\sqrt{\text{Hz}}$)$^{-1}$.

4.5 Experiment 1: Texture Identification

As described in [41], the transducer was used in a causality inversion process: recording force and stimulating with displacement, but instead of asking participants to simply compare the roughness of a virtual surface with that of a real one, we asked them to identify different textures, thereby showing that they could discriminate textured surfaces in the complete absence of stimulation distributed in space. The principle of the experiment was to first record interaction with five different texture samples. A group of participants were then asked to identify three of these virtual samples among the five real samples, and another group of participants to match three of the real samples with the five real samples, leaving much possibility for confusion. We expected participants to be able to identify the real or the

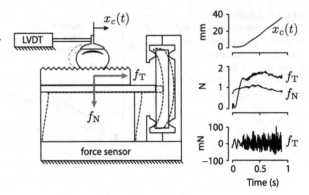

Fig. 4.9 Sensor operation. A textured sample is bonded to the central plate. The finger position is measured by a LVDT sensor and the total interaction force monitored by a six-axis force sensor

virtual textures with an equivalent level of performance. We also expected that a learning effect would be made apparent from the order of testing.

4.5.1 Methods and Materials

4.5.1.1 Design

We used a 5-alternative forced-choice matching procedure during which the participants were asked to identify a comparison stimulus with the five standard textures. With the first group of five participants, in a first session, the comparison stimulus was a real surface picked randomly in a set of three and the standard stimuli were the five real textures. They were then tested in a second session with a simulated surface picked randomly among the same set of three as the comparison stimulus and the standard stimuli were also the five real textures. A different group of five participants was also tested, but with the simulated stimuli first, and then with real comparisons. The two missing textures in the comparison set served as 'catch samples' to test the participants' ability to detect non matching textures. They also acted as distractors since the participants were looking for them.

4.5.1.2 Sensing Apparatus

The transducer was used to measure the interaction force of the author's finger sliding on a textured surface as in Fig. 4.9. The samples were placed on the tray of the transducer. The aforementioned strain-gauge based force sensor measured the low frequency components of the interaction force. The finger position was located by a precision LVDT sensor (SX 12N060, Sensorex SA, Saint-Julien-en-Genevois, France) attached to the fingernail.

The position of the finger, the net force, and the tangential force sensed by the transducer were recorded at 10 kHz, i.e. with sampling period $T = 10^{-4}$ s. The

Fig. 4.10 Stimulator
operation. **a** The transducer is
mounted on linear bearing.
The slider is located by an
encoder. **b** The control
voltage is updated of
piezoelectrical actuator as a
function of the position and
the texture profile used. The
open loop command used

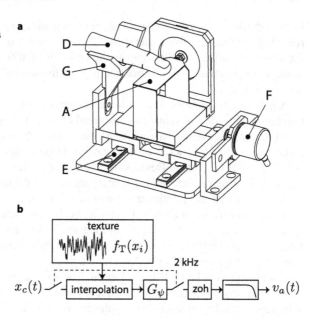

transducer signal was processed by computing its Fourier transform, truncating the
spectrum to 15–500 Hz and then reconstructing the signal. This method ensured
that the signal was restricted in the band where the transducer had a flat frequency
response, excluding any artifact and phase distortion.

4.5.1.3 Stimulating Apparatus

The friction force is correlated with the position of the finger, hence the signal is better expressed in the spatial domain and must be converted into this domain. Since
the speed of a finger is of the order of 50 mm s^{-1} the displacement resolution had
to be better than 5.0 μm. An interpolation procedure was employed to reconstruct
the signal in the spatial domain with the required resolution from the time-sampled
force, $f_T(jT)$, and from the time-sampled finger position $x_c(jT)$, $j \in \mathbb{N}$. The discrete functions $x_c(jT)$ and $f_T(jT)$ were first fitted with piecewise cubic Hermite
functions, $\mathscr{S}_x(t) \simeq x_c(t)$ and $\mathscr{S}_f(t) \simeq f_T(t)$, and then re-discretized. Given a space
sampling period, ε, position was resampled into samples $x_i = i\varepsilon$, $i \in \mathbb{N}$. The force
could then be represented in the discrete space domain following

$$f_T(x_i) = \mathscr{S}_f\big(\mathscr{S}_x^{-1}(x_i)\big). \tag{4.7}$$

The spatial sampling period was chosen to match the smallest step achievable with
the apparatus, $\varepsilon = 1.0$ μm.

The transducer was guided by a precision linear bearing, E, located by an encoder, F (Model R119 Gurley Precision Inc, Troy, NY, USA) that could resolve position with 1.0 μm precision. The fingertip rested on the tray, A, and the transducer

tracked the position of the proximal phalanx, D, resting in a cradle, G, see Fig. 4.10a, relieving the fingertip from lateral loads. As the slider moved with the finger, the transducer stimulated the fingertip as shown in Fig. 4.10b. Each 0.5 ms the position of the slider was read, the software interpolated the drive signal from a pre-recorded texture profile.

A gain G_ψ was adjusted to calibrate the stimulation for perceptually equivalent roughness with the real sample as described in [41]. To ensure that the drive signal matched the mechanical bandwidth of the transducer, an active analog 3th order Butterworth filter with a cut-off frequency 600 Hz (Salen–Key configuration with amplifier OP2177, Analog Devices Inc., Norwood, MA, USA). The filter served as a signal reconstruction filter and also compensated for the mechanical resonance.

Assuming that the Fourier transform of the texture signal exists, the accuracy of its restitution can be analyzed in terms of the maximum achievable spatial frequency for particular speeds, $\dot{x}_c(t)$, of the finger. The signal generated digitally was filtered at $v_{cut} = 500$ Hz. It had for effect to limit the largest reproducible spatial frequency to, $\mathscr{F}_{max} = v_{cut}/\dot{x}_c$ [7]. Assuming a finger velocity of 50 mm s^{-1}, the finest represented grating had a 0.1 mm period. The spatial resolution of the device was $\varepsilon = 10^{-3}$ mm, the largest spatial frequency was $\mathscr{F}_{max} = 10$ mm^{-1} so the textures were represented by at least 100 samples per period. At higher speeds, the temporal sampling became the limiting factor. In terms of the Nyquist sampling frequency, the fastest variations were reconstructed with at least $\dot{x}_c/(\mathscr{F}_{max}T) = 12$ temporal samples.

This analysis shows that the device could reproduce spatial grating as low as 0.1 mm, with high fidelity and a low sampling noise, both in spatial and temporal domain.

4.5.1.4 Test Bench

In order to ensure that the matching of virtual and real textures were performed under conditions that were as similar as possible, a test bench was constructed where the simulated and the real surfaces had to be touched under the same constraints, see Fig. 4.11a. In the two conditions, the user's finger was resting on a cradle that was attached to the stimulator. Since all moving parts were rigidly attached, participants felt the same inertia and the same mechanical imperfections in the two conditions. To ensure that the real and simulated surfaces felt thermally equivalent, a smooth PVC plate was glued to the tray of the stimulator.

4.5.1.5 Stimuli

Five textured samples were used as stimuli. They were 40 mm long, 20 mm wide, 3 mm thick, and made out of PVC plastic (see Fig. 4.12). The texture on each sample was created using different machining processes as described in Table 4.2. The resulting micro-geometries were not perceivable without relative motion. The interaction force resulting from scanning with a finger was acquired as described earlier. To

Fig. 4.11 Experimental setup. **a** The comparison stimuli bench. The stimulator and a real sample are side by side. During exploration, the participants experienced the stimuli under identical conditions. **b** Participants were asked to feel the comparison texture and to match it with the standard textures. They answered by pressing one of the five buttons

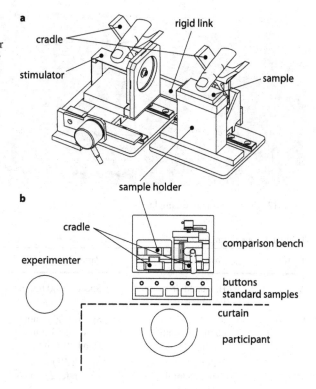

do so, one of the participants scanned the surfaces at a constant speed of 50 mm s^{-1} ±15 mm s^{-1} and a constant normal force of 0.5 ± 0.25 N to match the typical values of natural exploratory movements [32]. Before each measurement a solvent was used to clean the surfaces and that participants washed their hands. Recordings were resampled using the method described earlier.

4.5.1.6   Participants

Ten volunteers were recruited for the study. They were seven male and three female from the staff of CEA-LIST, aged 25 to 31. Three of them had experience with haptic devices, but all of them were naive as to the purpose of the experiment. They all were right-handed. They all gave their informed consent and did not report any motor or tactile deficit.

4.5.1.7   Procedure

Participants were placed in front of the setup hidden by a curtain, as illustrated in Fig. 4.11a. They were explained verbally and by a schematic the overall layout, without disclosing the details of the apparatus and then were explained their

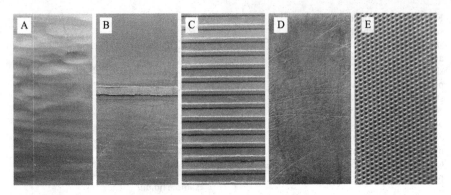

Fig. 4.12 Photos of the five samples

Table 4.2 Machining processes

Texture	Process	Parameters	Observations
A	Handheld drill Polishing disk	Speed: 6000 tr min^{-1}	Melted the surface
B	Milling 40 mm endmill Multiple teeth	Speed: 100 tr min^{-1} Feed rate: 1 mm s^{-1} Depth: 0.1 mm	1.5 mm ridge 0.1 mm high
C	Milling 2 mm endmill Two-flute	Speed: 5000 tr min^{-1} Feed rate: 2 mm s^{-1}	3 mm period 0.1 mm deep Grating
D	Coarse sandpaper	P40 grade	Random scratches
E	Drilling with 60° Conical end-mill	Depth: 0.1 mm 0.5 mm spacing	Quincunx pattern

task. They then donned acoustic isolation headphones (1015210, Sperian Protection, Roissy-CDG, France) that provided 30 dB of sound attenuation. White noise was also played through the headphones, and the volume could be adjusted to a comfortable level. There was a small light attached to the curtain to cue the participants to the next trial.

The five real textures samples were placed on a jig right behind the curtain and each was associated to a push-button, see Fig. 4.11b. Behind the answer panel was the comparison stimulus bench slightly elevated so that the jig did not perturb exploration. The experimenter guided the third phalanx of the participants to rest in the cradles. Three of the five textures, C, D, and E were used as comparison stimuli whether real or simulated.

Before each session, all five comparison stimuli, either real or simulated, were presented successively to the participant until they would be familiar with them. They typically became familiar with the stimuli after two rounds, but did not have to memorized them. A group of five participants were presented simulated textures first, and the other five real textures as comparison stimuli first. They were then all

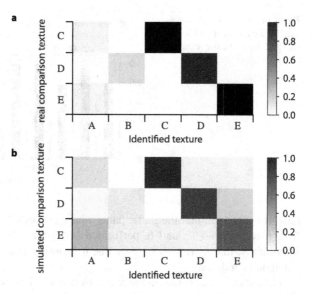

Fig. 4.13 Confusion matrices. **a** Matching real textures. **b** Matching simulated with real textures. The *gray scale* represent the proportion of correct answers

tested in the other condition in the second session. During the trials, the comparison stimuli C, D, and E were presently randomly. Participants identified the samples by matching them with one of the five standard textures.

When real textures were used as comparison, the experimenter manually changed the sample according to the instructions of the computer. The trials stopped whenever the session duration exceeded 15 minutes, typically after 30 or 40 trials, or when the trial number reached a hundred, whichever came first. In the case of virtual textures, since the process was faster, all participant performed 100 trials under 15 minutes. The volunteers were interviewed after each session to record their subjective experience.

4.5.2 Results

The overall results can be summarized by the confusion matrices shown in Fig. 4.13. The answer rates are shown by a gray scale from no match (white) to perfect match (black). When the real textures were used as comparison stimuli, Fig. 4.13a, identification was nearly perfect, which showed that it was possible to identify the samples. There was some confusion with the textures that were not used as comparison. After the experiment, the participants reported that they felt the need to detect them even though they were never presented. All noticed, however, that some samples were missing. When the comparison textures were simulated, Fig. 4.13b, the pattern was similar and the identification rate high but there was noticeably more confusion between samples A and E.

The rates of success of each participant in the two conditions are presented in Fig. 4.14. Although no detailed statistics were computed due to the small amount of

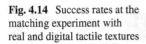

Fig. 4.14 Success rates at the matching experiment with real and digital tactile textures

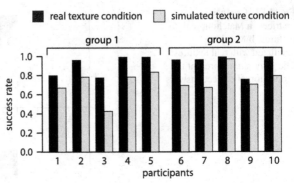

data, it is apparent that the participants who were tested with the real comparisons first, participants 4, 5, and 6, performed better than their counterparts, but more importantly, that there was transfer due to learning from the real condition to the simulated condition.

The overall success rate in the real texture condition was 0.93, with 0.10 of standard deviation. In the simulated texture condition the mean success rate was 0.75 with 0.14 of standard deviation among all participants. The group who performed the task in the real condition first had a success rate of 0.79 with standard deviation of 0.11 in the simulated condition.

Upon debriefing, we learned that all the participants, except one, felt a difference between the two sessions. They also described realistic sensations of rough textures, but the lack of an associated sensation of friction disturbed them somewhat. Six of them noticed the absence of relative motion between the stimulator and their finger. The confusion between sample A and E seems to be due to the fact that the perceived magnitudes of roughness were almost identical.

4.5.3 Discussion

The results suggest that the texture recognition task can be adequately performed but say nothing regarding the fidelity of the representation. Key indications regarding the perceptual accuracy of the sensations given by the apparatus concern the spectral properties of the stimuli in addition and their intensity. The ability of the apparatus to convey stimulation magnitude was already tested in [41], so we designed an experiment aimed at testing the participants' ability to discriminate the spatial frequency of pure tones.

4.6 Experiment 2: Tone Discrimination

We characterized the realism of the display by asking participants to discriminate pure spatial tones and by comparing the results with data found in the literature. The participants were asked to discriminate the frequency of a single sinusoidal grating that of a simulated counterpart. The Weber fraction, extracted from the data, was then compared to the results of [24].

4.6.1 Materials and Methods

4.6.1.1 Participants

Eight volunteers, 6 male and 2 female were recruited from the staff of CEA LIST (age 23 to 31). Two of them had experience with haptic interfaces, but were naive about the purpose of the study. They all were right-handed and did not report any somatosensory deficits. They verbally gave their inform consent.

4.6.1.2 Stimuli

The reference stimulus was a single sinusoidal wave epoxy grating accurately reproduced by a molding process of the very same 1.76 mm-spatial-period and 12.8 μm-amplitude grating used in [24]. The process used RTV silicon that can reproduce details as small as 1 μm. The first author's finger response was recorded as described in Experiment 1. The average finger speed was 64 mm/s and the normal force was 0.74 N. Comparison stimuli were delivered as previously described, with the difference that the spatial scale was stretched with a ratio r. The samples was then truncated and stiched to adjust the length of the records. The amplitudes of the virtual gratings were scaled by the factor r so the slope of the undulation would have the same value as described in [19]. Six stretching factor were tested: -70 %, -50 %, -30 %, -10 %, $+10$ %, $+30$ %, $+50$ % and $+70$ %, that is, the spatial period of the simulated textured varied from 0.53 mm to 2.99 mm.

4.6.1.3 Procedure

A 2-AFC constant stimuli procedure was carried out using the same bench as in Experiment 1, Fig. 4.11b. The standard stimulus was rigidly bonded to the support of the left of the bench. The comparison stimuli were presented as described earlier. Participants sat behind a curtain that concealed the apparatus and listened to the same white noise. They were ask to sense the texture by resting their finger in a cradle to experience the comparison and the standard stimulus alike. After a short instruction of the task, they could experience the standard stimulus and the comparison stimuli from the smallest spatial period to the largest twice. They were then presented randomly stretched samples and asked to tell which one of the two had the smallest spatial period.

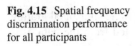

Fig. 4.15 Spatial frequency discrimination performance for all participants

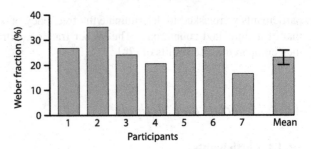

They gave their answer by keystrokes. The procedure stopped when all the samples were presented at least 10 times. The total procedure took 30 minutes at most.

4.6.1.4 Analysis

The data analysis described in [24] was carried out. The results were fitted cumulative Gaussian distribution using a Maximum Likelihood techniques. The 0 % stretching corresponds to the point of subjective equivalence, so its value should correspond to the 50 % proportion. We used a prior on the position of the inflection point of the statistic distribution at 0 % of stretching coefficient. Therefore, only the slope of the distribution was adjusted. The Weber fractions were retrieved from the stretching factor that led to the 75 % threshold. One of the participant, had abnormal results that are not reported.

4.6.2 Results and Discussion

Figure 4.15 shows the results for each participant. The average Weber fraction is 25.3 % with a standard deviation of 5.8 %. Using similar procedure but with real samples, Nefs et al. found a Weber fraction of 15.5 % [24].

While the discrimination performance using our apparatus caused a small deficit in performance compared to real samples, the results show that participants could discriminate spatial frequencies adequately.

4.7 Discussion: Implications for Human and Robotic Touch

This discussion follows directly from the above findings that were obtained from the performance of human participants. In fact, they apply also to robotic touch if robots of the future are to be endowed, like humans, with the faculty to detect, discriminate, and identify textured surfaces instantly. The discussion is organized as a set of observations.

4.7.1 The Absence of Spatially Distributed Information Does Not Imply Temporal Representation

The experimental conditions in which we placed the participants forced them to base their judgement on just two perceptual cues: stimulation magnitude and spectral content, since spatially distributed information was completely eliminated, as well as low frequency signal components and vertical movements. Yet they performed very well at matching textures and the majority reported a high degree of realism despite missing cues such as relative slip. This observation begs the question of the choice of domain in which spectral content of tactual textures should be represented, implicitly implying the domain in which textures might or should be processed.

One possibility is to represent the mechanical signal of interaction, the rapid skin lateral displacements specifically, in the time domain, like in acoustics. Several facts argue against this option. First, in contrast to acoustics, the representation would depend crucially on the condition under which the signal is acquired, namely on the scanning velocity, and hence would not be invariant, something which is perceptually troublesome. Secondly, while touch is capable of fine temporal discrimination (5–10 ms), unlike audition, its time-domain spectral processing abilities are poor [3, 11, 21]. In fact, when experiencing the signals that we have collected directly and without correlated movement is not felt by naive participants as texture but as what it is: vibrations. In fact, it is quite difficult to discriminate the textures tactually on the basis of temporal information only.

4.7.2 Candidate Representation: The Spatial Spectrograms and the Spatial Formant Organization

Another option is to represent texture in the space domain, even if it was acquired through regular time intervals, by expressing the interaction force as a function of finger position, that is, using a transformation such as that discussed in Sect. 4.5.1.3. An analysis in terms of the variation of spectral components through space gives rise to "spatial spectrograms" that express at each point in space the distribution of signal energy in terms of spatial frequencies: from 'smooth to sharp', rather than 'low to high' in the time domain.

The spatial spectrograms of the five standard texture where computed using a short-term fast Fourier transform with a 10 mm Blackman window and with zero-padding to match the length of the temporal representation. The results can be seen in Fig. 4.16 where each texture corresponds to a distinctive, highly structured pattern. This fact is quite intriguing. Recall from Fig. 4.1 that a texture profile, a triangular wave in this case, is converted through scanning into a complex interaction signal by a highly nonlinear transformation, despite the fact that the finger is a soft, deformable, low curvature object, but with high frequency details [30, 31]. After nonlinear transformation by the finger, the five textures are compactly represented by highly distinguishable structured patterns.

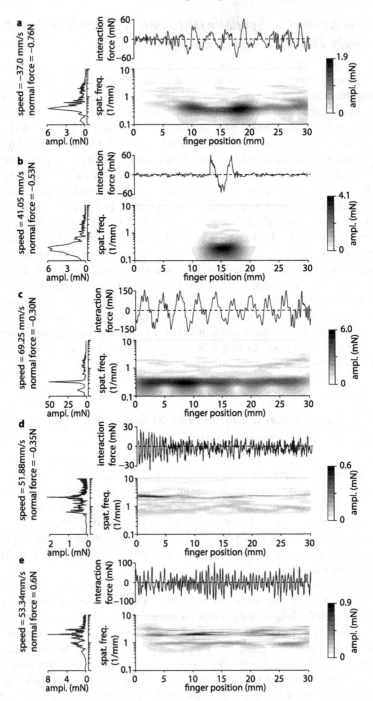

Fig. 4.16 Force-position profile along with the spatial spectrograms (expressed in spatial frequencies and position) and the average spectrum of each measurement made on the original textures

Each has a particular "formant" organization, to adopt a notion from acoustics, which are two-dimensional regions of high signal energy and that are not necessarily harmonic or even quasi-harmonic. The presence of this organization further argues against the notion of time-domain texture processing. For instance, in texture B, the original 1.5 mm-wide geometrical ridge is tactually represented by a round energy peak, 10 mm-wide and one-decade high. That texture samples A and E were hardly confused in the real condition but that in the simulated condition texture E was relatively frequently mistaken for A is interesting. The corresponding spatial spectrograms do share some common features but are shifted in spatial frequency. It could be that the very low frequencies of A contained spatially distributed information that was eliminated by the experimental conditions. Surprisingly, samples D and E were not confused frequently. Although their spatial formants differ, E is noticeably stronger than D, thus supporting a two-cue hypothesis. If one cue fails to provide reliable information, the other takes over.

4.7.3 Other Possibilities Based on Neurophysiology

Of course, other representations would be possible and with appropriate transformations could even be equivalent to the spatial spectrograms. Chiefly among them in the context of textures, are scalograms. These representations would also be compact and informative but have the inconvenience of depending on the arbitrary choice of a particular wavelet function—unless some optimality principle could be invoked. The short-time Fourier transform has the advantage that the only arbitrary parameter is the window length, which is worth discussing. The 10 mm window used in the short-term Fourier transform was selected simply because it is about the size of the contact surface of a scanning finger. In essence, this means that an isolated spatial feature should give signal inside a 10 mm window during scanning and be silent outside. Other window sizes could also be based on other optimality properties, for instance, be based the size of receptive fields of particular classes of mechanoreceptors in the skin [15, 37].

4.8 Conclusion

We designed an apparatus that was able to record with high precision and wide bandwidth the force of interaction between a finger and a textured surface. We first found by examining the results of scanning a "simple texture" that the interaction mechanics were complex and nonlinear that transform the underlying geometry into a broadband signal with little harmonic connection with the original geometry. We then used this apparatus to record different textures and inverted the process to reproduce as precisely as we could the original vibrations of the skin, but discarding distributed information and normal movements. Participants were still able to identify those textures with a high rate of success and several reported a keen yet imperfect experience of realism. Interestingly, the deficit of realism was not due, consciously, to the absence of distributed information or vertical movement but rather to the absence of the sensation of sliding friction, something we intend to correct in the near future.

For about a century, the notion that tactual texture perception is dependent on the relative sliding of a finger against a surface has been the subject of much discussion. Underlying this discussion is the assumption that the signals of interest to the sensing organism is a vibration pattern dependent on time, like in acoustics, combined with a spatial detection mechanism distributed on the skin. In our method we stimulated the finger with vibrations arising from a bare finger scanning naturalistic, textured surfaces, but dependent on space, that is, the stimulation depended on how the subject moved which they were free to choose, and there was no information regarding the distribution of stimulation on the skin. We then computed spatial spectrograms using a short-term Fourier transform with a 10 mm window and found that fingers transformed the original textures into a signal that could be represented as a spatial formant organization that compactly encoded the original surface. Our experiments are not incompatible with the hypothesis that these, or similar space-based transformations, could be employed by the nervous system to identify textures at perceptual speed. Such an approach might also be applicable to texture-aware robots.

Acknowledgement   The author would like to thank Edwige Pissaloux for insightful comments and Astrid Kappers for the loan of exquisitely made sinusoidal gratings. This work was supported by the Agence Nationale de la Recherche (ANR) through the REACTIVE project (ANR-07-TECSAN-020).

References

1. André T, Lefevre P, Thonnard JL (2010) Fingertip moisture is optimally modulated during object manipulation. J Neurophysiol 103(1):402–408
2. Bennett JM (1992) Recent developments in surface roughness characterization. Meas Sci Technol 3(12):1119–1127
3. Bensmaïa SJ, Hollins M (2000) Complex tactile waveform discrimination. J Acoust Soc Am 108(3):1236–1245
4. Bergmann-Tiest WM, Kappers AML (2006) Analysis of haptic perception of materials by multidimensional scaling and physical measurements of roughness and compressibility. Acta Psychol 121(1):1–20
5. Biet M, Giraud F, Lemaire-Semail B (2007) Squeeze film effect for the design of an ultrasonic tactile plate. IEEE Trans Ultrason Ferroelectr Freq Control 54(12):2678–2688
6. Bolanowski SJ, Gescheider GA, Verrillo RT, Checkosky CM (1988) Four channels mediate the mechanical aspects of touch. J Acoust Soc Am 84(5):1680–1684
7. Campion G, Hayward V (2005) Fundamental limits in the rendering of virtual haptic textures. In: Proceedings of the first joint eurohaptics conference and symposium on haptic interfaces for virtual environment and teleoperator systems, pp 263–270

8. Cascio CJ, Sathian K (2001) Temporal cues contribute to tactile perception of roughness. J Neurosci 21(14):5289–5296
9. Connor CE, Johnson KO (1992) Neural coding of tactile texture: comparison of spatial and temporal mechanisms for roughness perception. J Neurosci 12(9):3414
10. Cruz-Hernandez M, Hayward V (2001) Phase control approach to hysteresis reduction. IEEE Trans Control Syst Technol 9(1):17–26
11. Goff GD (1967) Differential discrimination of frequency of cutaneous mechanical vibration. J Exp Psychol 74(2):294–299
12. Heller MA (1982) Visual and tactual texture perception: intersensory cooperation. Percept Psychophys 31:339–344
13. Hollins M, Faldowski R, Rao S, Young F (1993) Perceptual dimensions of tactile surface texture: a multidimensional scaling analysis. Percept Psychophys 54(6):697–705
14. Krueger LE (1982) Tactual perception in historical perspective: David Katz's world of touch. In: Schiff W, Foulke E (eds) Tactual perception; a sourcebook. Cambridge University Press, Cambridge, pp 1–55
15. LaMotte RH, Whitehouse J (1986) Tactile detection of a dot on a smooth surface: peripheral neural events. J Neurophysiol 56:1109–1128
16. Lang J, Andrews S (2011) Measurement-based modeling of contact forces and textures for haptic rendering. IEEE Trans Vis Comput Graph 17(3):380–391
17. Lederman SJ (1983) Tactual roughness perception: spatial and temporal determinants. Can J Psychol 37(4):498–511
18. Libouton X, Barbier O, Plaghki L, Thonnard JL (2010) Tactile roughness discrimination threshold is unrelated to tactile spatial acuity. Behav Brain Res 208:473–478
19. Louw S, Kappers AML, Koenderink JJ (2000) Haptic detection thresholds of Gaussian profiles over the whole range of spatial scales. Exp Brain Res 132(3):369–374
20. Maeno T, Otokawa K, Konyo M (2006) Tactile display of surface texture by use of amplitude modulation of ultrasonic vibration. In: Proceedings of the IEEE ultrasonics symposium, pp 62–65
21. Mahns DA, Perkins NM, Sahai V, Robinson L, Rowe MJ (2005) Vibrotactile frequency discrimination in human hairy skin. J Neurophysiol 95(3):1442–1450
22. Millet G, Haliyo S, Regnier S, Hayward V (2009) The ultimate haptic device: first step. In: IEEE world haptics conference 2009, pp 273–278
23. Morley JW, Goodwin AW (1987) Sinusoidal movement of a grating across the monkey's fingerpad: temporal patterns of afferent fiber responses. J Neurosci 7:2181–2191
24. Nefs HT, Kappers AML, Koenderink JJ (2001) Amplitude and spatial-period discrimination in sinusoidal gratings by dynamic touch. Perception 30:1263–1274
25. Pai DK, Doel K, James DL, Lang J, Lloyd JE, Richmond JL, Yau SH (2001) Scanning physical interaction behavior of 3d objects. In: Proceedings of the 28th annual conference on computer graphics and interactive techniques, pp 87–96
26. Pataky TC, Latash ML, Zatsiorsky VM (2005) Viscoelastic response of the finger pad to incremental tangential displacements. J Biomech 38:1441–1449
27. Pawluk DTV, Howe RD (1999) Dynamic lumped element response of the human fingerpad. ASME J Biomech Eng 121:178–184
28. Persson BNJ (2001) Theory of rubber friction and contact mechanics. J Chem Phys 15(8):3840–3861
29. Picard D, Dacremont C, Valentin D, Giboreau A (2003) Perceptual dimensions of tactile textures. Acta Psychol 114(2):165–184
30. Scheibert J, Leurent S, Prevost A, Debregeas G (2009) The role of fingerprints in the coding of tactile information probed with a biomimetic sensor. Science 323:1503–1506
31. Smith AM, Chapman CE, Deslandes M, Langlais JS, Thibodeau MP (2002) Role of friction and tangential force variation in the subjective scaling of tactile roughness. Exp Brain Res 144(2):211–223
32. Smith AM, Gosselin G, Houde B (2002) Deployment of fingertip forces in tactile exploration. Exp Brain Res 147:209–218

33. Smits JG, Dalke SI, Cooney TK (1991) The constituent equations of piezoelectric bimorphs. Sens Actuators A, Phys 28:41–61
34. Stevens JC, Harris JR (1962) The scaling of subjective roughness and smoothness. J Exp Psychol 64:489–494
35. Takasaki M, Kotani H, Nara T, Mizuno T (2005) Transparent surface acoustic wave tactile display. In: Proceedings of the IEEE/RSJ international conference on intelligent robots and systems, pp 1115–1120
36. Thomas TR (1981) Characterization of surface roughness. Precis Eng 3(2):97–104
37. Vallbo ÅV, Johansson RS (1984) Properties of cutaneous mechanoreceptors in the human hand related to touch sensation. Hum Neurobiol 3:3–14
38. Vasudevan H, Manivannan M (2006) Recordable haptic textures. In: Proceedings of the IEEE international workshop on haptic audio visual environments and their application. HAVE 2006, pp 130–133
39. Wang Q, Hayward V (2007) In vivo biomechanics of the fingerpad skin under local tangential traction. J Biomech 40(4):851–860
40. Warman PH, Ennos AR (2009) Fingerprints are unlikely to increase the friction of primate fingerpads. J Exp Biol 212:2016–2022
41. Wiertlewski M, Lozada J, Pissaloux E, Hayward V (2010) Causality inversion in the reproduction of roughness. In: Kappers AML et al (eds) Proceedings of Europhaptics 2010. Lecture notes in computer science, vol 6192. Springer, Berlin, pp 17–24
42. Wiertlewski M, Hudin C, Hayward V (2011) On the $1/f$ noise and non-integer harmonic decay of the interaction of a finger sliding on flat and sinusoidal surfaces. In: World haptics conference (WHC), IEEE, New York, pp 25–30. doi:10.1109/WHC.2011.5945456
43. Wiertlewski M, Lozada J, Hayward V (2011) The spatial spectrum of tangential skin displacement can encode tactual texture. IEEE Trans Robot 27(3):461–472
44. Winfield L, Glassmire J, Colgate JE, Peshkin M (2007) T-PaD: tactile pattern display through variable friction reduction. In: World haptics 2007, pp 421–426
45. Yamamoto A, Nagasawa S, Yamamoto H, Higuchi T (2006) Electrostatic tactile display with thin film slider and its application to tactile telepresentation systems. IEEE Trans Vis Comput Graph 12(2):168–177
46. Yao HY, Hayward V, Ellis RE (2004) A tactile magnification instrument for minimally invasive surgery. In: Barillot C, Haynor DR, Hellier P (eds) Proceedings of MICCAI 2004. Lecture notes in computer science, vol 3217, pp 89–96

Chapter 5
Transducer for Mechanical Impedance Testing over a Wide Frequency Range

Abstract We describe a feedback-controlled active mechanical probe which can achieve a very low mechanical impedance, uniformly over a wide frequency range. The feedback produces a state of quasi-resonance which transforms the probe into a source of force used to excite an unknown load, resulting in a precise measurement of the real and imaginary components of the load impedance at any frequency. The instrument is applied to the determination of the mechanical impedance of a fingertip.

5.1 Opening Remarks

The signal calibration presented in Chap. 3, was performed assuming that the fingertip could be modeled by a simple elastic element. The next two chapters are devoted to investigating this assumption. This chapter describes a tunable mechanical probe with a view to test the mechanical impedance of the fingertip over a large bandwidth. The impedance measurement is performed by coupling the probe with the fingertip and by comparing the acceleration of the probe to the force applied to it. Since the impedance of the unloaded probe must be subtracted from the measurement to recover that of the load, the instrument is as sensitive as the probe's impedance is low. To enhance sensitivity, an closed-loop feedback was applied to decrease the impedance of the probe. Calibration of the apparatus and a preliminary measurement is reported.

5.2 Introduction

The present article describes an apparatus which is primarily intended for the measurement of the fingertip mechanical impedance, and of other objects of similar scale. The characterization of the bulk mechanical properties of the human finger plays an important role in the study of touch perception, in the design of tactile

---

Chapter is reprinted with kind permission from American Institute of Physics, originally published in [22].

M. Wiertlewski, *Reproduction of Tactual Textures*,
Springer Series on Touch and Haptic Systems, DOI 10.1007/978-1-4471-4841-8_5,
© Springer-Verlag London 2013

and haptic interface devices, in rehabilitation systems involving mechanical inter-action of the hand with surfaces, and in other fields such as the detection of skin pathological conditions [5, 9, 12–15, 21]. For instance, the study of the stability and robustness of the control of haptic devices depends on such knowledge [11]. Another example is in the area of tactile stimulators, where an accurate recording and reproduction of tactual signals depends on the knowledge of the mechanical characteristics of the skin [20, 23].

The concept of impedance and of its inverse—the concept of mobility—is useful to model and analyze the dynamics of mechanical, electrical, acoustic, hydraulic systems, and combinations thereof [8, 16]. In the mechanical domain, one considers the relationship between the force applied to an element and the resulting displace-ment. When linear, lumped analysis applies, an impedance can be represented as a combination of interconnected masses, springs, and dampers.

There are several approaches to measuring mechanical impedance. At the meso-scale, a widely used device is the so-called 'impedance head' employed in con-junction with an electrodynamic shaker. This device simultaneously records force and acceleration signals using two separate sensors. The inertial term resulting from the movements of the probing peg is subtracted from the force readings in order to access force and acceleration at the interaction point. Measurements involve acti-vating the shaker to excite the region or the object to be probed. Excitation can also be achieved though inertial forces, rather than from ground reaction [18]. Achieving collocated sensing, a prerequisite for accurate measurements, is difficult and colo-cation defects result in significant errors in the final impedance measurement [3]. Another approach is to recover the mechanical impedance from the variation of the electrical impedance of an electromagnetic transducer which can be measured ac-curately [6].

In nanotechnologies, the options are more limited. An approach to measuring mechanical properties at a very small scale is to use a vibrating cantilever driven at its natural resonance. The measurement then involves detecting amplitude changes for the same frequency (amplitude modulation) or the resonance frequency shift for the same amplitude (frequency modulation) of the tip of a cantilever in contact with a sample to be probed [1, 2, 10, 17]. This approach is particularly effective in vacuum since the Q-factor of the instrument, and hence its sensitivity, can reach large values. The displacement amplitude at resonance can then be much above the noise floor of the sensors, resulting in a high signal-to-noise ratio. Exciting a transducer at resonance is akin to reducing its impedance to a small value.

The instrument about to be described employs a resonance approach, yet, it is ca-pable of operating over a wide range of frequencies, instead of just one. The system is driven by a closed-loop controller that reduces the apparent impedance of an elec-tromagnetic transducer by almost an order of magnitude. Error propagation analysis shows how the feedback loop reduces measurement uncertainty. The impedance of the probed object can be recovered by subtraction of the unloaded response from the measurement made during testing.

The instrument is applied to produce frequency sweeps of the probing force to re-cover the impedance of a fingertip over a wide bandwidth. A complete measurement

Fig. 5.1 Principle of operation. The unknown load, z_u, perturbs the response of the probe

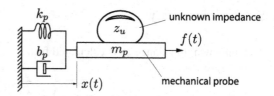

example is provided, revealing interesting properties of the mechanical behavior of the human fingertip.

5.3 Impedance Masking Approach

In the foregoing, a symbol in capital case designates the Fourier transform of a signal identified by the corresponding lower case symbol. Capital letters also denote the Laplace transform of a transfer function.

5.3.1 Principle

Referring to Fig. 5.1, the apparatus comprises a force generator, $f(t)$, acting on the probe associated to an impedance, z_p, having a tuned response. The response is affected by coupling the probe to an unknown load impedance, z_u, e.g. a finger. If the two impedances share the same velocity by mechanical connection, recovering the unknown impedance involves subtracting z_p from the total impedance z. In frequency domain we have,

$$Z_u(j\omega) = Z(j\omega) - Z_p(j\omega), \quad \forall \omega \in B,$$

where ω is the pulsation and B is the frequency range considered.

An impedance can be found by comparing the force applied, $F(j\omega)$, to the acceleration $\ddot{X}(j\omega)$, the velocity $\dot{X}(j\omega)$, and the displacement $X(j\omega)$,

$$Z(j\omega) \overset{\text{def}}{=} \frac{F}{\dot{X}}(j\omega) = j\omega\frac{F}{\ddot{X}}(j\omega) = -\frac{j}{\omega}\frac{F}{X}(j\omega).$$

If Z_p is known, the unknown load mechanics can be deduced at any frequency from the measured impedance using $b = \Re(Z_u(j\omega))$ and $m\omega - k/\omega = \Im(Z_u(j\omega))$.

This approach is effective only if the impedance of the probe is commensurate with or smaller than that of the load. In general, the smaller is z_p before z_u, the better is the measurement. In this article we describe a closed-loop control approach to reduce the impedance of the probe without suffering from sensor noise amplification.

5.3.2 Error Analysis

The unknown impedance z_u is calculated from

$$z_u = \frac{f}{\dot{x}} - \frac{f}{\dot{x}_p},$$

where F is the force output from a transducer, $\dot{X}$ the velocity of the probe coupled to the load, and $\dot{X}_p$ the velocity of the unloaded probe. Assuming white Gaussian noise in the sensors and a rigid probe, the variance of the measurement can be expressed in terms of individual components,

$$\sigma_z^2 = \left(\frac{\sigma_f^2}{f} + \frac{\sigma_{\dot{x}}^2}{\dot{x}}\right)\left(\frac{f}{\dot{x}}\right)^2 + \left(\frac{\sigma_f^2}{f} + \frac{\sigma_{\dot{x}_p}^2}{\dot{x}_p}\right)\left(\frac{f}{\dot{x}_p}\right)^2.$$

If σ_f and $\sigma_{\dot{x}}$ can be considered to be constant, it follows that in order to minimize the variance of σ_Z, the unloaded probe displacement should be as high as possible, i.e the probe mobility should be as high as possible,

$$\lim_{\dot{x}_p \to \infty} \sigma_z^2 = \left(\frac{\sigma_f^2}{f} + \frac{\sigma_{\dot{x}}^2}{\dot{x}}\right)\left(\frac{f}{\dot{x}}\right)^2.$$

This relationship is at the core of the resonant measurement principle. Low-loss transducers oscillating at resonance have an impedance that is close to zero, displacement is maximized, and the measurement of the load impedance is optimal. At this point, there are two possible paths to follow in order to achieve an active reduction of impedance for any frequency using a single transducer. They are discussed in the next section.

5.3.3 Active Feedback Control Approaches

A first approach would be to use feedback to construct a closed-loop system that behaves like a high-Q resonant system whose frequency can be placed arbitrarily. The resonant tuning approach is attractive, but is only applicable over a narrow range. Given a certain physical transducer, forcing the closed-loop system to resonate above the transducer's natural frequency becomes increasingly difficult with rising frequencies, since the input drive signal amplitude must also increase, leading to saturation.

Another inherent limitation comes from the sensitivity behavior of closed-loop system [7]. Given G, the system transfer function and C, the feedback, the magnitude of the sensitivity function, $|S| = 1/|(1 + GC)|$, cannot be kept low when $|G|$ becomes small, since there is no freedom in the choice of C. The result is an increasing sensitivity to parameter errors, precluding the achievement of high-Q closed-loop behavior in the high frequencies. Such an approach also limits the measurement options since identification techniques employing random signal excitation are precluded.

Fig. 5.2 Closed-loop control.
Position, velocity and
acceleration are fed back to
the transducer to modify the
apparent impedance

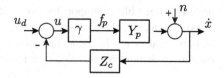

Another approach, adopted here, is to employ feedback to reduce the impedance of the closed-loop system over a targeted range of frequencies. If the system impedance can be kept uniformly low in this range, then many measurement options are possible, including the sine sweep excitation technique that is exemplified in this article. Intuitively, the objective is to maximize displacement by increasing the apparent mobility. The choice of the transducer is therefore critical. Laplace-force transducers or electrostatic force transducers have the desired natural, low-impedance characteristics. At the targeted length-scale (1–10 mm displacements), however, electromagnetic transducer, i.e. the voice-coil is the transducer of choice.

In the control diagram, Fig. 5.2, $Y_p = 1/Z_p$ represents the mobility of the probe, Z_c the impedance of the active feedback, γ the combined drive factor of the transducer and the gain of the amplifier, u the reference input, and i the current driven in the coil.

The output, $\dot{x}$, is a combination of sensor noise signal, n, injected in the loop and of the reference signal u_d,

$$\dot{x} = \frac{\gamma Y_p}{1 + \gamma Y_p Z_c} u_d + \frac{1}{1 + \gamma Y_p Z_c} n.$$

Assuming that the noise from the sensor is zero-mean, Gaussian with variance σ_n^2, and that the sensor is the dominant source of noise, the measurement variance is

$$\sigma_{\dot{x}}^2 = \left| \frac{1}{1 + \gamma Y_p Z_c} \right|^2 \sigma_n^2,$$

and the mean velocity is

$$\langle \dot{x} \rangle = \frac{\gamma Y_p}{1 + \gamma Y_p Z_c}.$$

The ratio $\sigma_{\dot{x}}/\langle \dot{x} \rangle$ does not change under closed-loop control, hence the measurement is not affected by closing the loop, as long as the feedback does not introduce additional significant errors. The control problem boils down to a pole placement problem to produce a uniform response over a range of frequencies, while achieving a reduction of the apparent impedance of the transducer at a value significantly smaller than that of the load to be measured.

5.4 Implementation

The concept was applied to the particular case of the measurement of the fingertip impedance. A device was constructed from a voice-coil motor driving a eight-bar flexural guide able to support the pressure of a finger. Since a voice-coil accurately transforms a current into a force, the sensors are expected to be the dominant source of noise and error. To avoid the need to design a state-observer, sensors directly measuring displacement, velocity and acceleration were included in the design.

5.4.1 Electro-Mechanical Arrangement

Referring to Fig. 5.3 a fingertip was constrained by a holder and was pressed against a surface which was guided by a flexure driven by a voice-coil motor (FRS8, VISATON GmbH, Haan, Germany). A force-sensor (Nano 17, ATI Industrial Automation, Apex, NC, USA) was placed under the ground link of the flexure to monitor the normal force component. It was also used to measure the tangential force component acting on the flexure for calibration the motor drive factor.

The flexure was of the eight-bar type which has the benefit of exact compensation of off-axis stresses and which therefore provides accurate linear guidance, even for large deflections [19]. It was cut out of acetal plastic.

The voice-coil was driven by a voltage-controlled current amplifier in order to compensate for the coil inductance and the back-EMF. The Laplace force generated by the coil was then proportional to the command voltage, u, see Fig. 5.4. The circuit was built from an operational amplifier (OPA548, Texas Instruments, Dallas, TX, USA) where the feedback was provided by a precision shunt resistance R_s. The transconductance gain was $i/u = -R_2/(R_1 R_s)$. Given a voice-coil with drive factor Bl, the total gain was

$$\gamma = \frac{f}{u} = -Bl\frac{R_2}{R_1 R_s}.$$

5.4.2 Impedance Feedback Control Loop

The objective was to reduce the apparent impedance of the system in order to maximize the difference between the unloaded and the loaded configurations. The control was obtained by feeding back position, velocity, and acceleration to the transducer through three gains that respectively modified the apparent stiffness, damping, and mass of the system. The open-loop transfer function in the Laplace domain was

$$\gamma u = (ms^2 + bs + k)x,$$

where m, b and k are the mass, damping coefficient and the stiffness of the actuator, and where s is the Laplace operator. In closed-loop operation, the position, x, the velocity, $\dot{x}$, and the acceleration, $\ddot{x}$, were fed back through gains l_k, l_b, and l_m, leading to

$$\gamma u = \left[ms^2 + bs + k - \gamma\left(l_m s^2 + l_b s + l_k\right)\right]x.$$

Fig. 5.3 Mechanical implementation of the impedance-meter

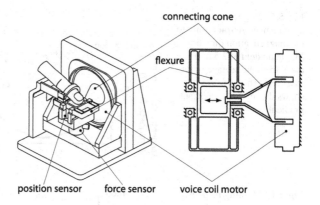

Fig. 5.4 Current-mode coil drive

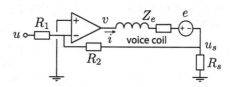

The apparent dynamic parameters were

$$\bar{m} = m - \gamma l_m, \qquad \bar{b} = b - \gamma l_b, \quad \text{and} \quad \bar{k} = k - \gamma l_k.$$

Thus,

$$Z(s) = \frac{\gamma u}{sx} = Z_p(s) - \gamma Z_c(s), \tag{5.1}$$

where $Z_c(s) = l_m s^2 + l_b s + l_k$, represents the controller impedance as in Fig. 5.2. Stability was ensured as long as the apparent dynamic parameters were strictly positive.

The feedback was implemented using analog circuits employing operational amplifiers (LMC660, Linear Technology Corp., Milpitas CA, USA) to compute the gains, sums, and differences that the control required.

5.4.3 Sensing

The position sensor was built from a hall-effect sensor (SS49, Honeywell, Morristown, NJ, USA) responding to the magnetic field of a semi-Halbach magnet configuration that created a uniform gradient over a large region. With three 5 mm cuboid neodymium-iron-boron magnets, a 4 mm region with a 0.1 T mm^{-1} gradient at a distance of 2.5 mm away from the magnets was achieved, see Fig. 5.5. A finite-element analysis showed good linearity over ± 2 mm range (linear regression with $R^2 > 99.9$ %). This configuration exhibited a five-fold advantage over the convention single-magnet arrangement. The noise floor of the position sensor was 2 µm.

Fig. 5.5 Magnet arrangement. *Arrows* point to the magnetization direction. Magnetic field at 2.5 mm away from the surface of the assembly. The *gray area* shows the region of constant gradient

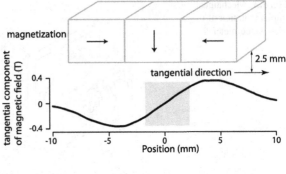

Fig. 5.6 Mobility response retrieved from the accelerometer (*thick gray*) and fitted model (*dashed*)

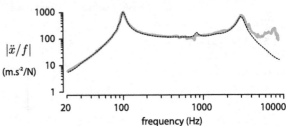

Velocity was measured from the back-EMF generated by the voice-coil. The voltage, v, across the coil terminals and the current, i, flowing through it (through the R_s shunt resistor) were measured. The circuit included the voice-coil electrical impedance, Z_e, in series with a voltage generator, $e = Bl\dot{x}$, and a known voltage generator, v. Kirchhoff's law gives $e = v - [(Z_e + R_s)/R_s]u_s$ from which $\dot{x}$ was easily derived with the analog electronics.

A commercially available accelerometer (2250A-10, Endevco, San Juan Capistrano, CA, USA) measured acceleration. Its mass was 0.4 g and its size was $5 \times 10 \times 3$ mm.

5.4.4 Control

The resulting system could be represented by a second-order system, but this approximation did not hold in the high frequencies. Since we aimed at wide bandwidth operation, the higher modes reduced or eliminated the stability margin at high gains. Autoregressive identification showed that the actual system could be well approximated by a 6-pole and 2-zero transfer-function from 20 Hz to 10 kHz. The lower two poles accounted for the second order behavior, and the remaining poles and zeros modeled a low-Q anti-resonance around 800 Hz and a sharper resonance at 3 kHz. The frequency response of the system and of the model are shown in Fig. 5.6.

From this model, the root loci for each feedback gain, acceleration, velocity, and position, were computed, see Fig. 5.7. Acceleration feedback decreased the apparent mass, but also reduces stability as the other poles moved toward the right-hand-side

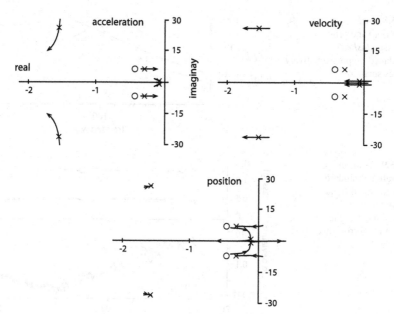

Fig. 5.7 Root loci. Effects of acceleration, velocity, and position feedback. *Arrows* show the effect of increasing gains. *Crosses* and *circles* are the poles and zeros of the open-loop transfer function, respectively

of the imaginary plane. Velocity feedback reduced the apparent damping. Stability was ensured only when the apparent damping was strictly positive. Position feedback modified the apparent stiffness. Stability was ensured as long as the apparent stiffness was strictly positive. With the aid of these diagrams, the system was tuned to achieve the response described in the next section.

5.5 Results

5.5.1 Unloaded Closed-Loop Response

The closed-loop frequency response in the targeted frequency band of the unloaded actuator is shown in Fig. 5.8, where it can be compared to the original open-loop response. With a good stability margin, the mobility was increased by a factor 5. The closed-loop accelerance was the same as the accelerance available in open-loop at the natural resonant frequency. A limiting factor of the present realization was the 800 Hz resonance. Performance could be increased in future realizations by optimization of the structural response of the suspension.

Fig. 5.8 Instrument response (*solid line*) and original response (*dashed*). The impedance is approximatively 5 times smaller

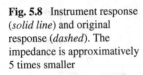
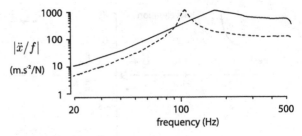

Fig. 5.9 Proof mass calibration. Standard deviation at all frequencies. Above 80 Hz the measurement of the accelerance error lower than 0.1 g

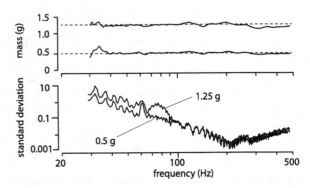

5.5.2 Proof Masses

Validation was performed using calibrated masses of 0.5 g and 1.25 g. Using sine sweep excitation for the measurements of the unloaded and loaded response, the impedance in 30–500 Hz band was retrieved, see Fig. 5.9. The standard variation was evaluated from 50 measurements. These measurements were extracted from the accelerometer signal which is the most accurate of the three sensors used in the system.

Above 100 Hz, measurement errors never exceeded 10 %. The standard deviation of the measurement follows the same amplitude pattern as the impedance and the relative acceleration was higher above 80 Hz. The uncertainty in the low frequency is caused by the low value of the masses impedance. Therefore measurements fall into noise.

5.5.3 Proof Cantilever

We fabricated a small elastic cantilever beam out of acetal plastic. Its response was measured independently from an impulsive test using a Laser Doppler vibrometer (OVF-2500 with OVF-534 head, Polytec Inc., Irvine, CA, USA). The tip of the cantilever was bonded to the moving plate of the apparatus using double sided tape and impedance measurements were performed in the 30–500 Hz bandwidth. The results can be seen in Fig. 5.10. The very low damping of the proof cantilever explain the

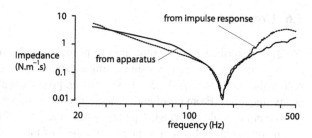

Fig. 5.10 Measurement of the proof beam. The impedance from the impulse response was obtained from displacement measurement divided by frequency

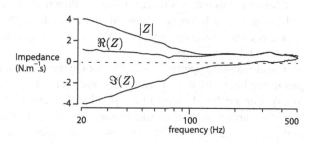

Fig. 5.11 Complete mechanical characterization of a fingertip

difference in the results in the high frequencies but the low frequencies the measurement followed the expected decrease of impedance with a rate of −20 dB/decade. The impedance value dropped around the resonant frequency. In the high frequencies, the impedance measurement showed the expected inertial behavior since the impedance increased at a rate of +20 dB/decade.

5.5.4 Fingertip Measurement

We can now show an example of a complete mechanical behavior measurement made while touching the probing plate of the instrument with a finger pushing on it with a normal force of 0.6 N, see Fig. 5.11. In the low frequencies, the probing displacement was of the order of one millimeter.

It can be seen from the real part of the measured impedance that the fingertip behaved essentially like a spring, up to a frequency of 100 Hz. Damping was significant and dominated above 150 Hz. The apparent mass of the fingertip was quite small, viz. 0.2 g, and its contribution to the response could be neglected within the range from DC to 500 Hz. Therefore, the finger could be modeled as a spring and a damper, with a transition at about 100 Hz.

These results are consistent with previous observations reporting that the fingertip skin can track unilateral stimuli up to about 100 Hz, albeit with normal excitation, [4] the cited study being the only one which, to our knowledge, has tested the fingertip skin behavior within the full frequency range that we can consider with our apparatus. Other measurements, not reported here, showed that the fingertip impedance varied significantly according to several factors.

5.6 Conclusion

We have described an apparatus able to probe the bulk mechanical impedance of a sample over a wide range, using a single actuator. A feedback loop was used to reduce the apparent impedance of the actuator, producing a state of quasi-resonance at any frequency. From proof masses and a proof cantilever, we could determine that the apparatus could detect a 0.1 g mass within the 20 to 500 Hz range.

The intended application is the measurement of the fingertip mechanical properties. Initial measurements revealed that the fingertip could be modeled by a spring and a damper—a Kelvin element—in the range from DC to 500 Hz, but that the impedance varied according to the testing conditions, a phenomenon which the subject of ongoing investigations.

The closed-loop operation principle for the reduction of apparent impedance, implemented here with a mesoscale electromagnetic device, could be easily extended to larger or smaller devices. In the small scales, electrostatic comb devices would scale favorably for actuation and sensing. Finally, state observers could be included in the control design to reduce the number of sensors, but their impact on the accuracy of the measurements would have to be investigated.

Acknowledgements   The author would like to thank José Lozada for his insightful comments and would also like to acknowledge the help of Mehdi Boukallel and Mathieu Grossard. This work was supported by the Agence Nationale de la Recherche (ANR) through the REACTIVE project (ANR-07-TECSAN-020). Additional funding was from the European Research Council, Advanced Grant PATCH, agreement No. 247300.

References

1. Acosta JC, Hwang G, Polesel-Maris J, Régnier S (2011) A tuning fork based wide range mechanical characterization tool with nanorobotic manipulators inside a scanning electron microscope. Rev Sci Instrum 82:035116
2. Asif SAS, Wahl K, Colton R (1999) Nanoindentation and contact stiffness measurement using force modulation with a capacitive load-displacement transducer. Rev Sci Instrum 70:2408
3. Brownjohn JMW, Steele GH, Cawley P, Adams RD (1980) Errors in mechanical impedance data obtained with impedance heads. J Sound Vib 73(3):461–468
4. Cohen JC, Makous JC, Bolanowski SJ (1999) Under which conditions do the skin and probe decouple during sinusoidal vibrations? Exp Brain Res 129:211–217
5. Dolan JM, Friedman MB, Nagurka ML (1993) Dynamic and loaded impedance components in the maintenance of human arm posture. IEEE Trans Syst Man Cybern 23(3):698–709
6. Doutres O, Dauchez N, Genevaux JM, Lemarquand G, Mezil S (2010) Ironless transducer for measuring the mechanical properties of porous materials. Rev Sci Instrum 81:055101
7. Doyle JC, Francis BA, Tannenbaum A (1990) Feedback control theory. Macmillan, London
8. Firestone FA (1933) A new analogy between mechanical and electrical systems. J Acoust Soc Am 4:249
9. Fu CY, Oliver M (2005) Direct measurement of index finger mechanical impedance at low force. In: Eurohaptics conference, 2005 and symposium on haptic interfaces for virtual environment and teleoperator systems, 2005. World haptics 2005. First joint, pp 657–659. IEEE, New York

10. Giessibl FJ (1997) Forces and frequency shifts in atomic-resolution dynamic-force microscopy. Phys Rev B 56(24):16010–16015. doi:10.1103/PhysRevB.56.16010
11. Gillespie B, Cutkosky M (1996) Stable user-specific rendering of the virtual wall. In: Proceedings of the ASME dynamic systems and control division. DSC, vol 58, pp 397–406
12. Hajian AZ, Howe RD (1997) Identification of the mechanical impedance at the human finger tip. J Biomech Eng 119:109
13. Israr A, Choi S, Tan HZ (2006) Detection threshold and mechanical impedance of the hand in a Pen-Hold posture. In: 2006 IEEE/RSJ international conference on intelligent robots and systems, pp 472–477. IEEE, New York
14. Israr A, Choi S, Tan HZ (2007) Mechanical impedance of the hand holding a spherical tool at threshold and suprathreshold stimulation levels. In: World haptics 2007, pp 56–60. IEEE, New York
15. Kuchenbecker KJ, Park JG, Niemeyer G (2003) Characterizing the human wrist for improved haptic interaction. In: Proc ASME int mechanical engineering Congress and exposition, vol 2, p 42017
16. O'Hara GJ (1967) Mechanical impedance and mobility concepts. J Acoust Soc Am 41:1180
17. Park SJ, Goodman MB, Pruitt BL (2007) Analysis of nematode mechanics by piezoresistive displacement clamp. Proc Natl Acad Sci 104(44):17376
18. Sands SA, Camilleri P, Breuer RA, Cambrell GK, Alfredson RJ, Skuza EM, Berger PJ, Wilkinson MH (2009) A novel method for the measurement of mechanical mobility. J Sound Vib 320(3):559–575
19. Smith ST, Chetwynd DG, Bowen DK (1987) Design and assessment of monolithic high precision translation mechanisms. J Phys E, Sci Instrum 20(8):977–983
20. Wang Q, Hayward V (2010) Biomechanically optimized distributed tactile transducer based on lateral skin deformation. Int J Robot Res 29(4):323–335
21. Wang Q, Kong L, Sprigle S, Hayward V (2006) Portable gage for pressure ulcer detection. In: Proceedings of the IEEE engineering in Medicine and Biology Society conference, pp 5997–6000
22. Wiertlewski M, Hayward V (2012) Transducer for mechanical impedance testing over a wide frequency range through active feedback. Rev Sci Instrum 83(2). doi:10.1063/1.3683572
23. Wiertlewski M, Lozada J, Pissaloux E, Hayward V (2010) Causality inversion in the reproduction of roughness. In: Kappers AML et al (eds) Proceedings of Eurohaptics 2010. Lecture notes in computer science, vol 6192. Springer, Berlin, pp 17–24

Chapter 6
Mechanical Behavior of the Fingertip in Lateral Traction

Abstract In the past, it was observed indirectly that the mechanical properties of the fingertip could be characterized by elasticity from DC to about 100 Hz and by viscosity above this frequency. Using a high mobility probe specifically designed to test the mechanical impedance of small viscoelastic objects, we measured accurately the impedance of the fingertip of seven participants under a variety of conditions relevant to purposeful touch. Direct measurements vindicated the previous observations that interactions forces could be explained by linear elasticity up to an average of 100 Hz and by linear viscosity beyond this frequency, while inertial contributions could be neglected. We also characterized the dependency of the fingertip impedance upon normal load, orientation, and time. We discuss the implications of these results with regard to microscopic nature of fingertip tissues and the tactual perception of textures.

6.1 Opening Remarks

As mentioned in the preface of Chap. 5, this chapter is aimed at the precise determination of the dynamic properties of the fingertip, over the 20–500 Hz frequency band. This determination is accomplished by means of the mechanical probe described in the previous chapter. The identification of mechanical impedance of the fingertip stimulation in the tangential direction was performed under several conditions. The study revealed that fingertips behaved elastically with a stiffness closed to 1 N mm^{-1} until a corner frequency of about 100 Hz, above which damping became dominant. A damping value of 2 N m^{-1} s could be attributed to the viscosity of subcutaneous tissues. Inertial terms were swamped by the dominant viscosity and elasticity. It was found that the effective equivalent inertia of the fingertip was of the order of 200 mg or less. This study vindicates earlier assumption but also suggests that tactile display, haptic displays and synthesis algorithms should account for finger viscosity when operating at high frequency.

Chapter is reprinted with kind permission from Elsevier, originally published in [31].

6.2 Introduction

There is indirect evidence that contact with the fingertip can be represented by a dominantly elastic load, in the small displacement range and up to a frequency of about 100 Hz. Above this corner frequency, the load presented by the fingertip becomes essentially viscous. This evidence was obtained by Cohen et al. [4, Fig. 6] using stroboscopic illumination in combination with an optical proximity detector to observe the movements of the glabrous skin when excited by a probe vibrating in the range 0.5 to 240 Hz and with displacements up to 1 mm. They found that the probe had a tendency to decouple from the fingertip skin for increasingly smaller probe displacements past a frequency of 80 Hz, an occurrence which is indicative of phase shift between force and displacement. Lamoré et al. [16] employed a clever paradigm to stimulate mechanoreceptors through two different excitation methods. One method was a conventional sinusoidal excitation and the other was through the amplitude modulation of a 2 kHz carrier. They found that the behavior of the skin could be represented by a high-pass mechanical filter with a corner frequency at 80 Hz [16, Fig. 4].

These two findings are mutually consistent when considering that the skin operates as a transmission medium in the latter experiment, whereas it acts as a load to the environment in the former. These findings are also in accordance with earlier results obtained from vibrating the skin of the arm or of the thigh [7]. Although the finger and these body areas differ anatomically, a probe contact area of 2.17 cm^2, which is commensurate with the contact area of a finger on a flat surface, gives a cross-over frequency of 100 Hz [18, Fig. 2].

We hypothesized that the fingertip could be well represented in the tangential direction by an elastic load up to about 80 Hz, and by a viscous load above this frequency. In addition we hypothesized that the inertial contribution to the response of a fingertip would be negligible over the entire frequency range relevant to touch. In order to test these hypotheses we developed a specific instrument, a mechanical probe achieving a very high mobility (having a stiffness smaller than 1.2 N m^{-1}, a damping smaller than 0.5 N m^{-1} s, and an accelerance smaller than 10^{-3} N m^{-1} s^2) which could closely approximate a pure source of force compared to the load of a fingertip.

With this instrument, we found that the tangential elasticity of the index fingertip of seven participants ranged from 0.6×10^3 to 2.0×10^3 N m^{-1}, a result that is inline with the results obtained by previous studies. Moreover, we found that the viscosity of the fingertip load ranged from 0.75 to 2.38 N m^{-1} s and that the inertia of the tissues entrained by a light touch ranged from 110 to 230 mg. These figures show that, indeed, viscous forces dominate the elastic and inertial contributions above 100 Hz.

Detailed test protocols were developed to evaluate the effects of normal load, stimulation orientation, and of time on these parameters. We also computed an estimate of the effective Young's modulus of the fingertip from contact mechanics considerations.

6.3 Previous Approaches to the Direct Determination of Fingertip Mechanics

The development of haptic display technology has motivated several studies related to the determination of the dynamics and statics of the human upper extremities. The study which is the closest to the present one is that of Hajian and Howe [8] who identified the mechanical impedance of the finger and found that a finger could be represented by a mass-spring-damper system. Their results refer to the behavior of an entire finger and were obtained through a impulsive test delivered by a pneumatic piston directed against the finger. The excitation contained little high-frequency energy and therefore the identification was reliable only in the low frequencies and where the finger underwent significant rigid-body displacement. Another single-finger study is that of Kern and Werthschützky [14] who used conditions that are more comparable to those of the present study. In these works, the measurement approach is that of the 'impedance head' whereby force and acceleration signals are simultaneously measured in the proximity of the interface between the probe and the sample, in order to infer the mechanical impedance of the unknown load. Such an approach is known to provide unreliable results in the high frequencies owing to a lack of confidence in truly co-located measurements [3].

Several other studies were performed in quasi-static conditions [12, 23, 24]. With specific reference to loading through lateral traction, Nakazawa et al. [19] modeled the fingertip as a spring and a damper, and found the values of $0.5\,\mathrm{N\,m^{-1}}$ and $2\,\mathrm{N\,m^{-1}\,s}$, for elasticity and viscosity respectively. Pataky et al. [21], using similar methods, modeled the elastic behavior of the fingertip and the relaxation effect and found values ranging from to $1\,\mathrm{N\,m^{-1}}$ and 11 s, for stiffness and relaxation duration.

The computational modeling of the fingertip pulp, which may be viewed as a membrane—the skin—filled a viscous liquid or gel—the subcutaneous tissues— [5, 25, 27, 28, 34] could be employed to predict the bulk response of a finger from elemental material properties [6, 20, 26, 29], but the results are dispersed due to the numerous assumptions that must be made to construct and solve the models, not mentioning the lack of anatomical realism that these models must contend with [9].

6.4 Materials and Methods

To perform direct, accurate measurements of the mechanical properties of a load, it is crucial to use an instrument, an excitation probe, that stores small amounts of energy compared to that circulating in or being dissipated by the sample. In other words, the probe should have a small mechanical impedance (or high mobility) compared to that measured. The converse approach, to maximize the impedance of the probe so that all the energy in the sample is reflected in the sample—the impedance head approach—has the shortcomings alluded to earlier. Other approaches involving, for instance air pressure modulation through jets or ultrasonic waves, would not do justice to the contact mechanics in effect during actual tactual behavior. For

Fig. 6.1 Impedance
measurement apparatus. The
tested finger was fixed in a
hinged cradle. An
electrodynamic transducer
drove a flexure guiding a
contact plate

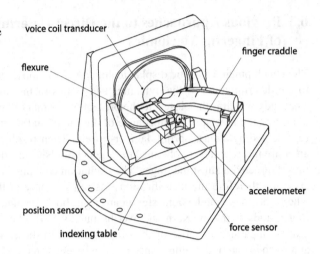

these reasons, we developed a mechanical probe, described in the next section, that possesses the required properties.

To ascertain the validity of the small-signal approximation we performed measurements at different excitation amplitudes and collected the results. Similarly, since the finger mechanics was expected to vary with the normal load, we performed series of measurements for various normal loads. Furthermore, since the mechanical properties of the fingertip were not expected to be isotropic, we performed measurements in different directions of stimulation. It is known that the finger exhibits viscoelastic properties operating at different time-scales. To ascertain this effect, we performed repeated measurements over periods of 20 s. With the results of measurements made under these conditions, we could paint a fairly complete picture of the response of a fingertip to tangential loading operative during purposeful touch.

6.4.1 Apparatus

The measurement apparatus, see Fig. 6.1, comprised a probing plate suspended by an eight-bar flexure that guided its movements in the tangential direction. It was stiff in the normal direction. The probe was driven by a voice-coil actuator. To reduce its impedance, and therefore enhance the signal-to-noise ratio of the instrument, an analog closed-loop feedback control was implemented.

Position, velocity, and acceleration measurements were fed back to the transducer as show in Fig. 6.2, where $f_d(t)$ is the desired force applied to the contact and $\dot{x}(t)$ its velocity. The system was tuned using the pole-placement approach that reduced the effective impedance of the system to a small value across the entire DC-500 Hz frequency range. Let $Y_p = 1/Z_p$ represent the mobility of the probe and Z_c be the impedance of the active feedback, the closed-loop apparent impedance of probe was $Z_a = (1 + Y_p Z_c)/Y_p$. The load impedance was acquired by comparing the unloaded probe response to the response when the finger is in contact.

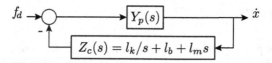

Fig. 6.2 Closed-loop control diagram of the probe implemented with analog electronics feeding back sensed position, velocity and acceleration

With this system, the impedance of the probe was always smaller than that of the load. In fact it was possible to consider the probe to be a pure force transducer over most of the frequency range, with the exception of the highest frequencies. The instrument was mounted to a fixed, strain-gauge force sensor (Nano 17, ATI Industrial Automation, Apex, NC, USA) that measured the averaged normal and tangential components of the interaction force for calibration purposes and contact condition monitoring. A detailed description of this apparatus is provided in [30].

The tested finger was immobilized by a strap onto a hinged cradle that could be adjusted around the last phalanx. The hinge allowed the participants to push freely on the probe surface. The finger was fixed relatively to the ground but the probe could be oriented by 15° steps by means of an indexing table that rotated about a vertical axis passing through the center probe plate.

The evaluation of the load impedance was made from two measurements. The input signal, f_d, was given the form of a frequency sweep and the output of the accelerometer, $\ddot{x}$, was recorded. Using the Fourier transform of the two signals (Welch method), F and $\ddot{X}$, the impedance Z was computed from

$$Z(j\omega) = j\omega \frac{F(j\omega)}{\ddot{X}(j\omega)},$$

where $j = \sqrt{-1}$ and ω was the angular pulsation. The impedance of the sample was obtained by subtracting the probe unloaded impedance from the coupled impedance.

6.4.2 Contact Condition

The contact surface was made from polycarbonate plastic and the surface finish was smooth. First, we experimented with bonding the fingertip to the plate using double-sided adhesive tape. We found that this condition created a dependency of the force-displacement curves upon loading or unloading the contact owing to a modification of the contact mechanics. We estimated that the coefficient friction, the ratio of the tangential interaction force component to the normal force component, $\mu = F_t/F_n$, was always higher than 1.2. Given this high value, we restricted the measurements conditions to values where there was no slip and did not introduce foreign elements at the interface. Moreover, we found that the force needed to detach a finger from the plate never exceeded 0.05 N, so that adhesion effects could be neglected.

6.4.3 Participants

Seven volunteers participated in the experiment, among which were four males and three females. They gave their informed consent. None of the participant reported any skin condition nor any injury to their fingers. The right index finger was measured in all participants. The participants' ages ranged from 23 to 32 with a mean of 25 years.

6.4.4 Procedure

The participants washed and dried their hands. Their index finger was fastened to the cradle. Before each experimental protocol, the impedance of the unloaded probe was calibrated by measuring it 10 times. Standard deviation was never above 1 % of the absolute value of the impedance and the mean value was used in the calculation of the fingertip impedance.

6.4.4.1 Small-Signal Linearity Protocol

The impedance of the finger pulp was measured for various tangential stimuli amplitudes. Participants were asked to remain as steady as possible and to regulate the normal force to 0.5 N via a visual feedback available from a computer screen. Once they could stabilize the normal force, the impedance was recorded during frequency sweeps lasting 1 s. Both proximal-distal and medial-lateral tests were performed. The signal amplitude varied from 0.0625 to 0.5 N by steps of 0.0625 N.

6.4.4.2 Normal Force Dependency Protocol

The relationship between the normal force and the impedance was evaluated using a 20–500 Hz frequency sweeps lasting 1 s with a standard amplitude of 0.25 N. The visual display of the normal force enabled the participants to control the normal force component and adjust it to a reference. One she or he could remain longer than 10 s within a 10 % tolerance range, and if the standard deviation of a measurement was smaller than 0.1 N, then the measurement was recorded and the reference force was changed. The normal force references values followed the sequence 0.25, 0.4, 0.5, 0.6, 0.75, 1.0; 1.25, 1.5, 2, 1.5, 1.0, 0.5, and 0.25 N.

6.4.4.3 Directional Dependency Protocol

The finger was tested with the same stimuli and in the same manner than in the normal force dependency protocol but the index table was used to vary the angle between the stimulation direction and the finger proximal-distal axis. The requested normal force component was 0.5 N and measurements were made each 15° for a total of 180° range. Impedances were recorded from both left to right and right to left direction then averaged in order to avoid drift of mechanical properties.

6.4.4.4 Time Dependency Protocol

This protocol aimed at investigating the change of impedance during steady finger pressure. In order to achieve higher temporal resolution, the length of the frequency sweep was reduced to 0.25 s and the bandwidth to 80–300 Hz. Participants stabilized the pushing force to remain within 10 % of an initial value of 0.5 N. Only medial-lateral stimulation was tested.

6.4.4.5 Young's Modulus Estimation

Th estimation of the effective Young's modulus from impedance measurements required the knowledge of the contact surface area. To this end, participants were ask to press their right index finger on a sponge filled with ink and then leave their fingerprints on a sheet of paper set on a scale. They pressed down slowly until reaching a normal force component of 0.5 N. They repeated this procedure four times.

6.4.5 Data Processing

6.4.5.1 Lumped Parameters Determination

The mechanical impedance of the fingertip in tangential traction was modeled by a mass-spring-damper system. In the frequency domain the mechanical impedance is expressed by

$$Z(j\omega) = b + j\left(m\omega - \frac{k}{\omega}\right)$$

where ω is the angular frequency, m is the moving mass, b is the viscosity coefficient, and k is the stiffness. The coefficient b was estimated from the real part of the impedance averaged over the frequency range. Stiffness and mass were determined using non-linear curve fitting on the imaginary part of the impedance. This procedure gave good estimates of stiffness and damping but it is apparent that a small moving mass cannot be accurately obtained from the above expression. It is far more effective to find the frequency for which the imaginary part of the impedance crosses zero and then to use the known stiffness to estimate the mass from $m = k/\omega^2$.

6.4.5.2 Effective Young's Modulus

The bulk stiffness and damping was the consequence of the deformation of the fingertip. Mechanical parameters of the material were estimated by assuming that the finger could be approximated by a sphere in contact with a flat, rigid plane. Further assuming that the material was linear, isotropic and homogeneous, contact mechanics allowed us to recover a value of an effective, averaged Young's modulus that could explain the measured behavior. Then, assuming first-order Kelvin–Voigt viscoelasticity of the fingertip, the complex modulus of the fingertip could be estimated.

For low tangential loads, it was fair to consider that the fingertip did not slip at any point of the contact surface and that the displacement was uniform inside the totality of the contact area. In polar coordinates, from [13] the no-slip condition leads to a distribution of traction that follows $q(r) = q_0(1 - r^2/a^2)^{-1/2}$ where a is the contact surface radius and where $q_0 = Q/(2\pi a^2)$ is the average traction. Deformation is given by the Boussinesq and Cerruti integral of a distributed tangential traction on a elastic half plane with the restriction $r < a$,

$$\delta = \frac{1}{2\pi G} \iint_S q(\xi, \lambda) \left(\frac{1-v}{\rho} + v \frac{(\xi - x)^2}{\rho^3} \right) d\xi \, d\lambda,$$

with $\rho^2 = (\xi - x)^2 + (\lambda - y)^2$, v the Poisson coefficient, and G the shear modulus. Young's modulus was found from $G = E/[2(v+1)]$. The evaluation of the above integral with the traction profile, $q_0(1 - r^2/a^2)^{-1/2}$, leads to an expression for the tangential stiffness,

$$K = \frac{Q}{\delta} = 8a \frac{G}{2-v},$$

from which E could be extracted given a and v (taken to be equal to 0.5 for soft tissues).

To capture the dynamic behavior into the previous model, we followed [17] in assuming that the stress can be expressed by the history of deformation. In the case of the Kelvin–Voigt model it follows that

$$\sigma(t) = E \varepsilon(t) + \eta \frac{d\varepsilon(t)}{dt},$$

where σ is the stress, ε is the strain and η is the viscosity. When driving the material with an oscillatory motion, the stress and strain follow $\sigma(t) = \bar{\sigma} e^{j\omega t}$ and $\varepsilon(t) = \bar{\varepsilon} e^{j\omega t - \phi}$, respectively. It is then possible to represent the material by a complex modulus

$$E^* = \frac{\sigma(t)}{\varepsilon(t)} = E + \eta j \omega.$$

Complex moduli were extracted from the impedance measurements using $K^*(\omega) = Z(\omega) j\omega$ where $K^* = k + bj\omega$ is the complex tangential stiffness derived from the expression of stiffness in combination with the complex modulus.

Fig. 6.3 Typical measurement. The real part represents damping. The imaginary part represents elasticity and inertia. The fingertip exhibits a second-order filter characteristic with dominant damping (*dashed lines*)

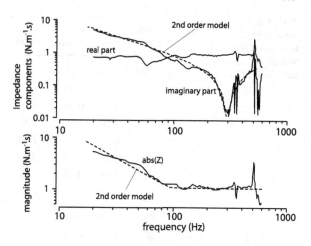

6.4.5.3 Contact Surface Area

The fingerprint marks were imaged using a flat bed scanner. An intensity threshold was used to removed the background and to isolate the prints. Morphological operations on the image filled the holes and cleaned isolated pixels. At this stage of the processing, the image was a blob representing the total area of contact. Ellipses were fitted to recover the major and minor axis and their half averages were used as a measure of the radius of the contact surface.

6.5 Results

6.5.1 Frequency Response

A representative example of the frequency response of the fingertip is plotted in Fig. 6.3.

The results show that the behavior of fingertips can be approximated by a highly damped second order system. From DC to approximatively 100 Hz, fingertips behave like an elastic spring. Beyond this frequency, damping is dominant. This results justify the adoption of a Kelvin–Voigt model for the fingertip in many applications. The imaginary part of the impedance exhibits a clear resonance at 300 Hz indicative of a transition from an elastic regime to an inertial regime.

The measurements frequently exhibited high frequency modes (400–500 Hz) in the imaginary part of the response. These higher modes had a fleeting character. They appeared and vanished seemingly randomly. They probably corresponded to the establishment and destruction of standing wave patterns in the skin according to certain, precise contact conditions. They were not investigated further.

Table 6.1 Dynamic parameters of the fingertips, Kelvin–Voigt model for a normal force component magnitude of 0.5 N

Part.	a mm	k N/mm	b N s/m	m g	E kPa	η Pa s	R^2 %
Medial-lateral direction							
cr	4.36	0.92	1.52	0.17	118	196	85
el	4.95	0.94	1.39	0.08	107	158	89
gt	4.80	1.67	2.38	0.16	195	279	85
lb	4.82	0.59	1.00	0.11	69	117	92
ma	4.26	0.75	0.80	0.08	98	106	93
mw	4.50	0.74	1.26	0.17	92	157	84
ss	4.40	0.78	0.94	0.12	98	119	90
Proximal-distal direction							
cr	6.27	1.84	1.99	0.23	164	179	88
el	6.65	1.52	1.53	0.17	127	128	87
gt	6.14	2.08	2.09	0.23	191	191	88
lb	7.40	1.10	1.19	0.14	83	90	83
ma	5.38	0.71	0.75	0.07	74	78	84
mw	6.20	1.56	1.64	0.18	131	149	87
ss	6.60	1.23	1.25	0.14	98	106	90

6.5.2 Mechanical Parameters

Table 6.1 summarizes the fitted mechanical parameters for all participants as well as the extracted effective elastic and viscous moduli.

One-way ANOVAs revealed that stiffness, damping, and mass were dependent on gender ($p = 0.026$, $p = 0.001$ and $p = 0.0021$ respectively). The same dependance is found for elasticity and viscosity ($p < 0.01$ in both cases). Stiffness did show a dependency on direction of stimulation ($p = 0.035$ whereas $p = 0.55$ and $p = 0.37$ for damping and inertia, respectively). Elasticity and viscosity failed to rejected the null hypothesis of a correlation with direction ($p = 0.27$ and $p = 0.36$).

6.5.3 Small-Signal Linearity

Stiffness and damping values were found for varying excitation force amplitudes (from 0.1 N to 0.8 N). From these values and the force amplitudes the position-force and velocity-force relationships were reconstructed as shown in Fig. 6.4. The actual force on the fingertip was measured to be in average 60 % of the excitation force, the remaining being lost in the self-impedance of the probe. The Spearman correlation coefficient, ρ, between stiffness and amplitude for each trials was always higher than 0.87 except for one outlier (gt) who obtained $\rho = 0.31$, on a scale of 0 to 1. Damping

Fig. 6.4 Plots of mechanical parameters values for all participants

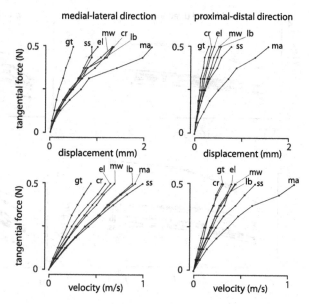

was also correlated with amplitude since the Spearman correlation coefficient was higher than 0.92 for all participants, also with an outlier (gt) for whom ρ had a value of 0.49. Linear fitting the position-force and the velocity-force characteristics led to a goodness of fit of $R^2 = 93 \pm 4$ % for stiffness and $R^2 = 97 \pm 2$ % for damping.

At high amplitudes, however, it is apparent that stiffness and damping had a tendency to decrease. It is unlikely that the decrement was due to changes in the mechanical properties. It is plausible that at higher amplitudes, partial slip may have taken place. In fact for highest amplitudes, the coefficient of static friction would have to be of about 1.6 to ensure adhesion where preliminary tests shown that the value of 1.2 was a reliable lower-bounds.

For the range of values relevant to touch (lateral deflections smaller than 2 mm, lateral component of force smaller than 0.5 N, normal component of 0.5 N) the measurements strongly support the conclusion that the finger may be considered to behave linearly in the bulk.

6.5.4 Normal Force

The normal component of force had impact on all three dynamic parameters. Their evolution as a function of normal force can be seen in Fig. 6.5 for all participants. Non-parametric Spearman correlation performed between the force and the dynamic parameter, gave a minimum value of 0.87 for stiffness, 0.88 for damping, and with the exception of three outliers, inertia produced correlations greater than 0.89.

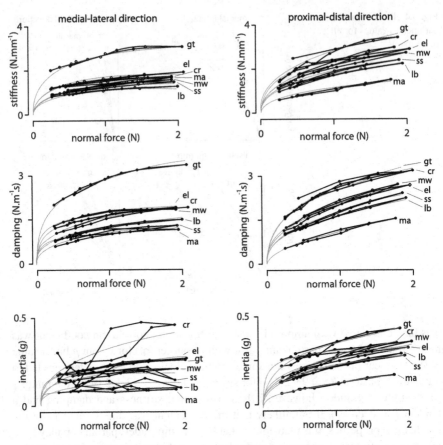

Fig. 6.5 From left to right, stiffness, damping, and inertia, respectively, as a function of normal force for all participants

Table 6.2 Power-law coefficients for the stiffness, damping and inertia dependency on normal force	α	β	R^2
Stiffness [N mm^{-1}]	0.35 ± 0.09	1.48 ± 0.60	96 ± 4.5
Damping [N m^{-1} s]	0.35 ± 0.10	1.78 ± 0.60	96 ± 3.6
Mass [g]	0.26 ± 0.19	0.20 ± 0.06	71 ± 20

The relationships were fitted with zero-intercept power-law regressions of the form, βP^{α}, where α and β are coefficients and P is the normal force component. Mean values and standard deviations of the coefficients as well as the goodnesses of fit are summarized in Table 6.2. The α coefficients are close to $\frac{1}{3}$, which consistent with the prediction made by Hertzian contact theory.

Fig. 6.6 Stiffness [N mm^{-1}] (*black lines*), damping [N m^{-1} s] (*grey lines*), and inertia [g] (*thinner black lines*) as a function of testing direction angle for all participants

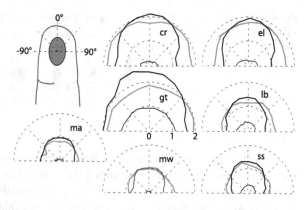

Fig. 6.7 Representative examples for participants gt and lb of the temporal evolution of stiffness and damping. The *gray lines* show linear regressions

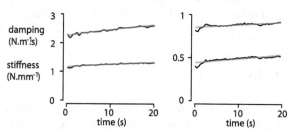

6.5.5 Orientation

Polar plots of the dependency of the mechanical parameters on orientation for all participants can be inspected in Fig. 6.6. It is evident that no two fingers are alike. For instance, subject ma has a low-stiffness, low-damping, low-inertia fingertip, whereas subject gt exhibits very different values. Participants lb, cr, el, and ss showed sharp direction tuning of the fingertip stiffness whereas other have a more uniform elasticity. We performed a one-way ANOVA in order to test the influence of stimulation. Stiffness significantly depended on orientation ($p = 0.054$) but the null hypothesis of dependency of damping and inertia on orientation was rejected ($p = 0.95$ and $p = 0.99$ respectively).

6.5.6 Time

Here, in order to enhance the temporal resolution of the measurements, the high frequencies were not tested. Therefore, the effects of inertia were not visible in the impedance measurements and cannot be reported. Representative results for two participants can be seen in Fig. 6.7. Figure 6.7 shows the rates-of-change of stiffness and damping for each participant when pushing down with a normal force of 0.5 N for 20 s.

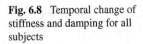 **Fig. 6.8** Temporal change of stiffness and damping for all subjects

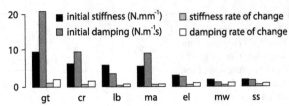

Figure 6.8 shows, for all subjects, how the mechanical parameters drifted through time. The rates were obtained by fitting linear regressions through the measurements. Because the slopes were shallow, the R^2 measure did not represent the goodness of fit well. The hypothesis of a correlation of the dynamics properties with time failed to be rejected ($10^{-10} < p < 0.055$), however.

6.6 Discussion

Using an instrument featuring very high mobility (stiffness lower $1.5\,\mathrm{N\,m^{-1}}$, damping lower than $0.5\,\mathrm{N\,m^{-1}}$ s, and inertia smaller than 1 g) we measured with high reliability the mechanical impedance of the index fingertip of seven participants under a number of loading conditions. We found that fingertips behaved like elastic springs up to a frequency of about 100 Hz, that the damping term dominated the interaction force beyond this corner frequency, and that the inertial term could be neglected in the range from DC to 500 Hz.

In some participants, the medial-lateral stiffness could be half of the proximal-distal stiffness but in other participants the two values were similar. In all participants, however, damping and inertia did not exhibit marked orientation selectivity. Under average loading conditions, the fingertip has a lateral stiffness of about $0.8\,\mathrm{N\,mm^{-1}}$ and a damping coefficient of about $1.2\,\mathrm{N\,m^{-1}}$ s, but these figures can vary greatly among people. When loaded, the mechanical parameters drift upward at a rate of about 0.5 % per second. The corner frequency of $(1/2\pi)(800/1.2) \simeq 100$ Hz vindicates the observations of Cohen et al. [4] and Lamoré et al. [16].

Assuming uniform bulk mechanics of the fingertip, the effective Young's modulus and viscosity were extracted from contact mechanics considerations. The Young's modulus mean values, 114.9 kPa in medial-lateral direction and 137.5 kPa in the proximal-distal direction, were found to be much lower than previously reported in other testing conditions. Wang and Hayward [29] found a average value of 2.5 MPa for the skin only. Pan et al. [20] found an average Young's modulus of 458 kPa in the forearm skin. Gennisson et al. [6] found an average of 2.5 MPa for the skin whereas the epidermis has an elasticity of 10 kPa. These differences are easy to explain, considering that a fingertip is not a homogeneous medium but a complex, fibrous, multi-phase structure. It therefore risky to extrapolate bulk properties from the characteristics of individual components. The structural organization of the fingertip therefore gives to this organ surprising compliance to external loading and make it appear like a viscous load at higher frequencies.

Finite element studies implied that the fingertip could resonate at 100–125 Hz [34]. We did not find any evidence of such phenomenon, quite the opposite was the case, owing to the high value of the damping coefficient. Viscous forces swamped out inertial forces over the entire frequency range. The values that we found are quite inline with those reported earlier. Nakazawa et al. [19] found an average damping coefficient in lateral displacement of 2.0 Nm^{-1} s during force-step testing to be compared with the 1.6 Nm^{-1} s average found in our study. Viscosity extracted from the contact mechanics considerations is also of the same order than those found with other techniques. The large amount of dissipation can be attributed to fluid displacement in the skin and in subcutaneous tissues [11].

When pushing the fingertip onto a flat surface, the area of contact grows rapidly [22, 25] such that it almost reaches its final value even at low normal loading. We found that the dependence of stiffness, damping, and inertia on the contact surface area, by-and-large, followed a $\frac{1}{3}$ power-law, which can be related to the size of the normal force component in Hertz's contact theory. From the identification of the small equivalent mass that is entrained by fast oscillatory traction of a fingertip, viz. 100 mg, we can surmise that only a very small amount of tissue stores and releases kinetic energy during contact. In other words, only must the superficial layer of the fingertip oscillate macroscopically during contact (a 1 mm thick over a 1 cm^2 contact surface gives 100 mg). The remaining energy must correspond to waves that dissipate in the body.

Our results could be applied to the interpretation of findings related to the perception of tactual textures, which are felt when sliding the fingertips over irregular surfaces. It is well accepted that for texture appreciation, the central nervous system relies on temporal information, i.e. transitory and persisting vibrations generated in the fingertips, at the detriment of spatial information, i.e. strain distributions in the fingertip [10, 15].

We have mentioned earlier that considering the fingertip to be a simple viscoelastic solid was an oversimplification of the actual physics. Recent studies have shown that the sliding interaction of a fingertip with flat, smoothly undulating, and textured surfaces engendered oscillations having energy in the whole frequency range [32, 33]. A high-level of bulk viscosity would be more consistent with the behavior of a bi-phasic solid, that is to say, a porous medium. While the mechanics of these media are notoriously difficult to model, it is know that beyond a characteristic frequency, they tend to behave as dispersive medium [1, 2]. In other words, the fingertip can be viewed as a mechanical filter that attenuates the high spatial frequencies when the temporal frequencies become high, yet high frequency transmission of temporal information is preserved.

Acknowledgements This work was supported by the French research agency through the REAC-TIVE project (ANR-07-TECSAN-020). Additional funding was provided by the European Research Council, Advanced Grant PATCH, agreement No. 247300.

References

1. Biot MA (1956) Theory of propagation of elastic waves in a fluid-saturated porous solid. I. Low-frequency range. J Acoust Soc Am 28(2):168–178
2. Biot MA (1956) Theory of propagation of elastic waves in a fluid-saturated porous solid. II. Higher frequency range. J Acoust Soc Am 28(2):179–191
3. Brownjohn JMW, Steele GH, Cawley P, Adams RD (1980) Errors in mechanical impedance data obtained with impedance heads. J Sound Vib 73(3):461–468
4. Cohen JC, Makous JC, Bolanowski SJ (1999) Under which conditions do the skin and probe decouple during sinusoidal vibrations? Exp Brain Res 129:211–217
5. Dandekar K, Raju BI, Srinivasan MA (2003) 3-d finite-element models of human and monkey fingertips to investigate the mechanics of tactile sense. J Biomech Eng 125:682
6. Gennisson JL, Baldeweck T, Tanter M, Catheline S, Fink M, Sandrin L, Cornillon C, Querleux B (2004) Assessment of elastic parameters of human skin using dynamic elastography. IEEE Trans Ultrason Ferroelectr Freq Control 51(8):980–989
7. von Gierker HE, Oestreicher HK, Francke EK, Parrack HO, von Wittern WW (1952) Physics of vibrations in living tissues. J Appl Physiol 4(12):886–900
8. Hajian AZ, Howe RD (1997) Identification of the mechanical impedance at the human finger tip. J Biomech Eng 119:109
9. Hauck RM, Camp L, Ehrlich HP, Saggers GC, Banducci DR, Graham WP (2003) Pulp non-fiction: microscopic anatomy of the digital pulp space. Plast Reconstr Surg 113:536–539
10. Hollins M, Bensmaia SJ (2007) The coding of roughness. Can J Exp Psychol 61(3):184–195
11. Jamison CE, Marangoni RD, Glaser AA (1968) Viscoelastic properties of soft tissue by discrete model characterization. J Biomech 1(1):33–36
12. Jindrich DL, Zhou Y, Becker T, Dennerlein JT (2003) Non-linear viscoelastic models predict fingertip pulp force-displacement characteristics during voluntary tapping. J Biomech 36(4):497–503
13. Johnson KL (1987) Contact mechanics. Cambridge University Press, Cambridge
14. Kern TA, Werthschützky R (2008) Studies of the mechanical impedance of the index finger in multiple dimensions. In: Ferre M (ed) EuroHaptics 2008. Lecture notes in computer science, vol 5024. Springer, Berlin, pp 175–180
15. Klatzky RL, Lederman SJ (1999) Tactile roughness perception with a rigid link interposed between skin and surface. Percept Psychophys 61(4):591–607
16. Lamoré PJJ, Muijser H, Keemink CJ (1986) Envelope detection of amplitude-modulated high-frequency sinusoidal signals-by skin mechanoreceptors. J Acoust Soc Am 79(4):1082–1085
17. Lee EH (1956) Stress analysis in viscoelastic materials. J Appl Phys 27(7):665–672
18. Moore TJ (1970) A survey of the mechanical characteristics of skin and tissue in response to vibratory stimulation. IEEE Trans Man-Mach Syst 11(1):79–84
19. Nakazawa N, Ikeura R, Inooka H (2000) Characteristics of human fingertips in the shearing direction. Biol Cybern 82(3):207–214
20. Pan L, Zan L, Foster FS (1997) In vivo high frequency ultrasound assessment of skin elasticity. In: Ultrasonics symposium, 1997. Proceedings, vol 2. IEEE, New York, pp 1087–1091
21. Pataky TC, Latash ML, Zatsiorsky VM (2005) Viscoelastic response of the finger pad to incremental tangential displacements. J Biomech 38:1441–1449
22. Pawluk DTV, Howe RD (1999) Dynamic contact of the human fingerpad against a flat surface. J Biomech Eng 121:605
23. Pawluk DTV, Howe RD (1999) Dynamic lumped element response of the human fingerpad. ASME J Biomech Eng 121:178–184
24. Serina ER, Mote CD, Rempel D (1997) Force response of the fingertip pulp to repeated compression—effects of loading rate, loading angle and anthropometry. J Biomech 30(10):1035–1040
25. Serina ER, Mockensturm E, Mote CD, Rempel D (1998) A structural model of the forced compression of the fingertip pulp. J Biomech 31(7):639–646

26. Silver FH, Freeman JW, DeVore D (2001) Viscoelastic properties of human skin and processed dermis. Skin Res Technol 7(1):18–23
27. Srinivasan MA (1989) Surface deflection of primate fingertip under line load. J Biomech 22(4):343–349
28. Tada M, Pai DK (2008) Finger shell: predicting finger pad deformation under line loading. In: Proceedings of the 2008 symposium on haptic interfaces for virtual environment and teleoperator systems. IEEE Computer Society, Los Alamitos, pp 107–112
29. Wang Q, Hayward V (2007) In vivo biomechanics of the fingerpad skin under local tangential traction. J Biomech 40(4):851–860
30. Wiertlewski M, Hayward V (2011) Transducer for mechanical impedance testing over a wide frequency range. TBD
31. Wiertlewski M, Hayward V (2012) Mechanical behavior of the fingertip in the range of frequencies and displacements relevant to touch. J Biomech 45(11):1869–1874. doi:10.1016/j.jbiomech.2012.05.045
32. Wiertlewski M, Hudin C, Hayward V (2011) On the $1/f$ noise and non-integer harmonic decay of the interaction of a finger sliding on flat and sinusoidal surfaces. In: World haptics conference (WHC). IEEE, New York, pp 25–30. doi:10.1109/WHC.2011.5945456
33. Wiertlewski M, Lozada J, Hayward V (2011) The spatial spectrum of tangential skin displacement can encode tactual texture. IEEE Trans Robot 27(3):461–472
34. Wu JZ, Welcome DE, Krajnak K, Dong RG (2007) Finite element analysis of the penetrations of shear and normal vibrations into the soft tissues in a fingertip. Med Eng Phys 29(6):718–727

Chapter 7
Vibrations of a Finger Sliding on Flat and Wavy Surfaces

Abstract The fluctuations of the frictional force that arise from the stroke of a finger against flat and sinusoidal surfaces were studied. We used a custom-made, high-resolution friction force sensor able to resolve milli-newton forces, we recorded those fluctuations as well the net, low-frequency components of the interaction force. Measurement showed that the fluctuations of the sliding force were highly unsteady. Despite their randomness, force spectra averages revealed regularities. With a smooth, flat, but not mirror-finish, surface the background noise followed a $1/f$ trend. Recordings made with pure-tone sinusoidal gratings revealed complexities in the interaction between a finger and a surface. The fundamental frequency was driven by the periodicity of the gratings and harmonics followed a non-integer power-law decay that suggested strong nonlinearities in the fingertip interaction. The results are consistent with the existence of a multiplicity of simultaneous and rapid stick-slip relaxation oscillations. Results have implications for high fidelity haptic rendering and biotribology.

7.1 Opening Remarks

This chapter focuses on the vibrations generated by a finger sliding on a surface. The study is motivated by the need to create algorithms that can synthesize virtual textures. We recorded the interaction force between a finger and six different surfaces made of smooth epoxy. The first surface was flat, and the vibrations generated by the friction with the finger revealed the presence of a background noise that followed a $1/f$ trend. The other five surfaces were sinusoidal gratings with varying wavelength and amplitude. The analysis of the spatial spectra of the force produced by sliding made the nonlinearities of the interaction quite evident. The resulting data could be used to design texture synthesizers using, for instance, waveshapping techniques.

Chapter is reprinted with kind permission from IEEE, originally published in [26].

M. Wiertlewski, *Reproduction of Tactual Textures*,
Springer Series on Touch and Haptic Systems, DOI 10.1007/978-1-4471-4841-8_7,
© Springer-Verlag London 2013

7.2 Introduction

When sliding on most surfaces, human fingers generate audible noise. If the surface in question is able to acoustically radiate—as in the case of the sounding board of a guitar—the noise can actually be quite loud, denoting that a significant portion of the work done during sliding is transformed into acoustic energy which is dissipated in part in the tissues and in the solid object. Although originating from the skin it has recently been found that these vibrations can propagate far, and over a wide frequency range, in the tissues of the arm [8]. The reader can easily verify that such noise is generated even when sliding a finger on the mirror finish of a glass surface, unless the contact is lubricated.

Since the early works of Katz [13], as well as with more recent studies [3, 10–12, 20], it is generally accepted that these vibrations play a determinant role in the perception of textures, or absence thereof. One factor behind the generation of the oscillations by the sliding skin is the distribution and nature of asperities on surfaces which result from erosion, wear, manufacturing, or growth of the material. Another factor is the nature of the material (or fabric) of which the surface is made, since two surfaces having the same geometry can create very different acoustic signatures. The finger also contributes peculiar geometrical and material properties that determine the tribology of the contact [1, 21].

The motivation for our research is in the area of haptic virtual reality and computer simulations, where tactual texture plays an important role in realism [5, 6, 22]. Interestingly, the synthesis of rich, artificial tactual textures have tended to resemble audio rendering techniques, being based on generating a spatial waveform that is played at the speed of exploration [22]. The artificial waveform can be created either from measurements [9], or procedurally from stochastic or fractal models [7, 19]. All these synthesis processes are based on the assumption that the contact is made by a rigid probe that obeys an elastic linear law, and on simplified solid friction models to generate stimuli. As previously presented, these assumptions break down completely in the case of a bare finger [27].

Almost no work has been dedicated to this analysis of the dynamics of finger-surface interactions during steady sliding, with the perspective of simulating it. The fingertip exhibits non trivial friction phenomena that are not well understood [1], yet even under the assumption of linear elasticity, the frictional properties of soft material with rigid, rough surfaces are not trivial to model [18]. The output of a measurement depends on many parameters such as the viscoelasticity and the non-linearity of the finger and the skin [16, 17, 23]. The friction force itself varies according to various conditions including normal loading and hydration of the skin [2, 24].

In the present study, we asked whether the vibrations generated during the slip of a finger on smooth and sinusoidal surfaces could be characterized in terms of their spatial spectrum content. In other words, we wondered whether these vibrations presented characteristics of invariance or regularities with respect to the applied force and the slip velocity, that is to say, whether these vibrations encode the underlying surface geometry.

Employing a specially engineered sensor [25], we recorded the force *fluctuations* of a finger sliding on accurately manufactured flat surfaces and sinusoidal gratings.

This sensor can resolve sub milli-newton forces with a high dynamic range. We then performed a spectral analysis of the tangential force component in the spatial domain and sought to discover invariants. The analysis of the spatial frequency content of these vibrations gave some insights into the mechanisms that can possibly be at play.

The friction force signal can be represented by the superposition of two contributions. One being a background noise that follows a $1/f$ relationship, and the second being, on average only, a harmonic expansion of the original frequency that decays according to a non integer power-law, suggesting a relaxation oscillatory process. Results have implications for the synthesis of artificial tactile stimuli and in the field of biotribology.

7.3 Choice of Surfaces

The motivation for the choice of sinusoidal profiles was not that they represent "pure tones" in the spectral domain, but rather that they provide a smooth change of curvature within the contact regions with the skin. In effect, in the range of amplitudes, periods and interaction force magnitudes that we studied, a finger is not guaranteed to interact through one single connected region and can possibly touch only at the apices of the individual ridges. We did not control for this possibility, but from a general perceptive, such occurrences are part of the normal tactual exploration process and were left intact. In any case, sinusoidal profiles guaranteed that the curvature varied smoothly within restricted intervals.

Other profiles that were considered include triangular and square profiles, all having first-order discontinuities to which the skin would be exposed. For the purposes of the present study, discontinuities have the grave inconvenience that sharp edges translate into undefined (infinite) local strains. Physically, undefined strains correspond to damage, a type of interaction which can include abrasion, delamination and other effects, and which are interactions that are normally avoided during tactual slip.

In the choice and manufacturing of the surfaces, we also considered the question of length scales, which is central to the concept of roughness. An undulating or a flat surface is compatible with the presence of irregularities at length scales that are 2 or 3 orders of magnitude smaller that the main textural components, the so-called 'surface finish' in plain language. Even if, due to the St Venant's principle, small-scale asperities are of no mechanical consequences a few microns inside the skin during static touch, such is not the case during sliding because their presence or absence has impact on skin tribology. There is a variety of phenomena at play including the possible presence of trapped foreign bodies and liquids that modify tribology according to the nature of small-scale asperities.

Healthy fingers permanently exude sweat and the *stratum corneum*, specifically, is a highly hydrophilic milieu. The presence of small-scale asperities, in contrast to amorphous glassy finishes, have impact on skin keratin plasticization and on the behavior of interfacial water, modifying adhesion [1].

Fig. 7.1 Measurement
Bench. The texture is
double-sided taped to the
piezoelectric force transducer
that measures frictional force
arising from the finger sliding
on it. The finger is located by
a LVDT fixed to the finger
through a universal joint

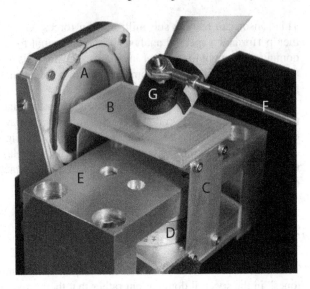

7.4 Finger-Surface Interaction Measurement

From the above considerations, to approach the question of the vibrations induced
by sliding, we performed highly accurate measurements of the tangential force component of the interaction force, that due to friction, during steady sliding of the finger
on flat and undulating sinusoidal surfaces with controlled surface finishes.

The apparatus comprised a finger position measurement device, a specifically
designed friction force transducer, and a conventional six-axis force sensor for a
complete characterization of low frequency interaction force components.

7.4.1 Hardware Components

The three hardware components measured three different aspects of the interaction,
namely: the position of the finger, the friction force fluctuations, and the net normal
and tangential forces. Sliding velocity was estimated from suitably processed differ-
entiation of the position signal. Figure 7.1 shows a picture of the apparatus during a
measurement.

The main component is a custom-made piezoelectric sensor that operates in the
direction tangential to the scanned surface in the 2.5–350 Hz frequency range and
with a sensitivity of $13\ \mathrm{N\,V^{-1}}$. The sensor was engineered to perform with noise
floor of 50 μN. The friction-force fluctuation transducer was built around a disk
multilayer piezo bender, A (CMBR07, Noliac Group A/S, Kvistgaard, Denmark) that
converted the tangential load force into electrical charges. The piezoelectric sensor
acted as a voltage generator coupled in series with a capacitor. The voltage was con-
ditioned by an instrumentation amplifier (LT1789, Linear Technology Corp., Milpitas, CA, USA). Two resistances passed the generated charges to the ground and

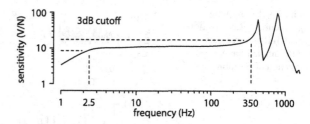

Fig. 7.2 Sensitivity of the sensor as a function of frequency from its impulse response. The flat band is the range 2.5 Hz to 350 Hz

therefore create a high-pass filter. The value of the resistances were chosen to produce a first-order, high-pass filter response with a 2.5 Hz cutoff frequency.

The interchangeable gratings, B, were bonded to a plate with double sided tape. The plate was suspended between two leaf springs, C, connected to the hollow center of the piezoelectric bender with a flexural joint that was very stiff in the axial direction. The resulting stiffness was 8.0×10^4 N m^{-1}, which is two orders of magnitude stiffer than a finger [16]. Hence, it provides a non-ambiguous measurement of the tangential sliding force. The suspended mass was estimated to be 12 g and therefore the natural frequency of the sensor was at 410 Hz. Figure 7.2 shows the response of the sensor derived from its impulse response.

The second element of the apparatus was a conventional strain gauge force sensor, D (Mini 40, ATI Industrial Automation, Apex, NC, USA) which responded to the low frequency force components along the normal and tangential directions. It was placed underneath the piezoelectric transducer on the load path to the mechanical ground, E. In order to maximize bandwidth, the mechanical arrangement, see Fig. 7.1, minimized the distance between the point of application of finger interaction force and the sensor flanges by mounting it 'upside-down'. The resolution was about 50 mN.

The subject's finger was located by a LVDT transducer (SX 12N060, Sensorex SA, Saint-Julien-en-Genevois, France) that could resolve 3 μm on a 40 mm stroke. Prior calibration was performed using precision micro-stage. The finger was rigidly linked to the movable core of the linear transducer, F, by a clip mounted on the nail side of the finger via a joint, G, totaling 5 degrees of freedom. The alignment of the LVDT was provided by adjustable linear stages.

Analog signals were acquired by a 16-bit data acquisition board (PCI-6229, National Instruments Corp., Austin, TX, USA) hosted by standard microcomputer. The sampling period was set at 10 kHz in order to have a comfortable sampling margin and reduce distortion during post-filtering.

7.4.2 Digital Signal Processing

The raw data were sampled in the temporal domain and therefore depended on the sliding velocity. In order to express the data with the velocity as a parameter, it had to be resampled in the spatial domain using a fixed spatial sampling period. The data retrieved from the measurement of a finger sliding on a given texture, Fig. 7.3a, were

Fig. 7.3 Resampling
procedure. **a** Raw position
data. **b** Raw force data.
c Resampled force as a
function of position

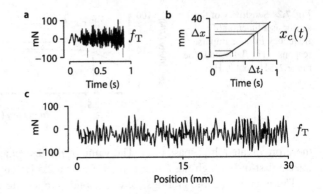

interpolated at each multiple of the discrete position steps obtained from the position trajectory as in Fig. 7.3b. The steady slip region was cropped out by truncating the initial transitory period of the recordings such that the samples were kept in the range 5 mm $< x_c <$ 35 mm. The interpolation-resampling procedure was applied to the tangential force measurements, f_T, from the piezoelectric transducer and to the normal and the tangential forces measured from the strain-gauge sensor.

The force frequency components above 350 Hz were filtered out of the measurements by performing a fast Fourier transform on the signal, truncating the spectrum and then reconstructing the signal using the inverse Fourier transform. The spectral domain processing provides a sharp cut-off, phase distortion-free signal conditioning. Velocity was computed by differentiating the position after applying a zero-phase moving average filter over 9 samples. It was then interpolated in space domain using the same procedure as above.

7.4.3 Gratings

We used sinusoidal gratings with height profiles expressed by $h(x) = A \sin(2\pi/\lambda)$. One profile, termed the nominal profile, had an amplitude of $A = 25$ μm and a spatial period of $\lambda = 1.96$ mm. The other four had different amplitudes, $A = 12.8$ and $A = 50$ μm, and spatial periods, $\lambda = 1.76$ and 2.5 mm.

The gratings were cast from the exact same stainless steel gratings that were employed in the study in reference [15]. We first made silicon molds (RTV 181, Esprit Composite, Paris, France) and then duplicates were manufactured by casting epoxy resin (EP 141, Esprit Composite, Paris, France). This process can reproduce details as small as 1 μm. Perceptually, the resulting epoxy duplicate gratings feel as rough as their steel originals. Microscopic inspection did not reveal any surface imperfections. They were milled down to 50 × 30 mm rectangular samples.

A flat surface was made with the same epoxy resin. The surface was rectified with a milling machine with a very low feed rate and a high speed. The resulting roughness was homogeneous with an Ra smaller than 1 μm and the surface finish felt similar to that of the sinusoidal surfaces.

7.4.4 Participants and Procedure

Two participants made the recordings. One of them was the first author. They were aged 24 and 26 and did not present scars or burn marks on their fingertips. They washed their hands before the session. They practiced the skill of maintaining the normal force and the velocity as constant as possible during recordings.

They sat in front of the apparatus such that their arms could move without resistance and recorded the sliding interaction force. Velocity and normal force were held as constant as possible during a single recording. The trials were selected according to how steady was the velocity and the normal force. Each participant made 50 satisfactory recordings with each texture. Recordings lasted less than one second, so that slowly varying effects due to viscoelastic behavior of a fingertip would not interfere with the results.

The velocity and normal force during every measurement matched standard exploratory conditions. Velocities ranged from $14 \, \mathrm{mm \, s^{-1}}$ to $339 \, \mathrm{mm \, s^{-1}}$ with a mean of $127 \, \mathrm{mm \, s^{-1}}$. Normal forces were in the range of $0.14 \, \mathrm{N}$ to $2.44 \, \mathrm{N}$ with a mean of $0.8 \, \mathrm{N}$.

7.5 Observations on the Results

The frequency content of the measurements made on the texture reveals multiple features of the transformation between the single wave grating and the engendered force fluctuations. Each of the following section discusses a feature of the signal. The first observation is that two recordings made in similar conditions do not produce the same spectral content. Nevertheless some properties emerge from the *averages* of large numbers of trials. Signal energy decays in a $1/f$ fashion and all harmonics are present with decreasing amplitudes according to a non-integer power law.

7.5.1 Lack of Stationarity

Even with similar interaction forces and velocities in, recordings presented great variations of amplitude and frequency content. Figure 7.4a, b, c shows the spectra of three measurements taken in similar conditions. In the low frequencies, the spatial spectra can have different decays and the amplitude of the fundamental changed. Its frequency is fixed, however, since it corresponds to that of the underlying surface. Harmonics are all present in most, but not all, of the recordings and their amplitudes also varied in each recording.

Averaging the spectra of many recordings, however, as shown in Fig. 7.4d, smooth the variations of spectra and a pattern emerges. The overall spectral content appears to be composed of two major components. The first component is a background noise that follows a $1/f$ law. The second component is, on average only, a periodic component with fundamental and harmonics. This $1/f$ noise and

Fig. 7.4 Three typical measurements with a finger sliding on a 50 μm amplitude, 1.96 mm spatial period grating. One hundred measurements *in grey*, and their average spectrum *in black*

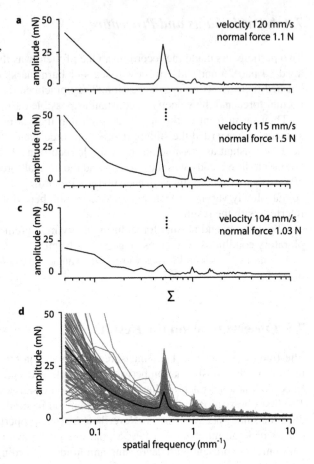

harmonics appear in the recordings in a non-predictable fashion, and not all the measurements exhibit these properties.

As one can expect, the fundamental frequency component is well represented and corresponded to the spatial frequency of the single-wave grating used. In addition to this fundamental, all harmonics appeared, suggesting the presence of strong nonlinearities in the transformation.

7.5.2 Energy Decay in Background Noise

Using recordings made with the flat surface, we further investigated the pattern of background noise. Figure 7.5 shows each individual measurement in light gray and the average spectrum in black. Furthermore, the spectrum of each individual measurement are fitted with a power function $S(k) = \beta/k^{\alpha}$ with $k = 1/\lambda$ being the

Fig. 7.5 Individual measurements spectra (*light gray*) and average (*black*) of the measurement made with the smooth flat surface

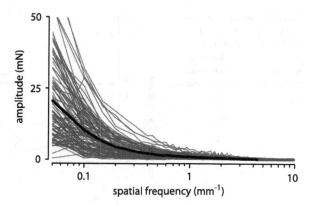

Fig. 7.6 Distribution of fitted coefficients α and β on the 100 measurements made with the smooth flat surface

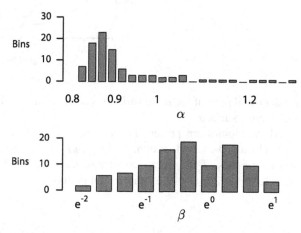

spatial frequency, α and β the fitting coefficients. The distribution of the fitted coefficients can be seen in Fig. 7.6.

The distribution of α coefficients reaches a maximum at 0.9. This feature indicates that the background noise is a fractal noise, found in numerous natural and man-made processes [4, 14]. The β value represents the magnitude of the fluctuations. This parameter is not related in any simple manner to any other parameter. The reader should keep in mind, however, that this noise behavior might be different for other materials than smooth epoxy.

The goodness of the fit for the power law is better than $R^2 = 0.93$ for most cases. Fifteen outliers out of a hundred have a goodness of fit around 0.8 and in one case the fit has a value of 0.3.

7.5.3 Fundamental and Harmonic Amplitudes

As it might be expected, averaging the measurements made on the three gratings with same amplitude $A = 25$ μm but varying spatial frequencies shows that the

Fig. 7.7 Spectrum average on the 3 spatial variations of the 25 μm gratings in *plain black*. Power law fitting on the amplitude of fundamental and its harmonics is plotted in *dashed grey*

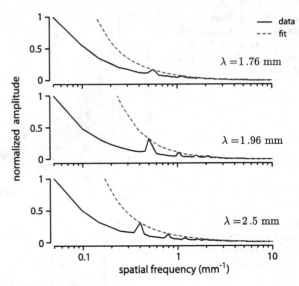

fundamental peak of the force variations always coincided with spatial frequency of the original surface.

All harmonics are present in all three cases. Their decay can be well represented by a power-law function, $S_h(k) = \beta_h / k^{\alpha_h}$. Fitting the data shown in Fig. 7.7 gives estimates of the decay coefficient $\alpha_h = \{1.5, 1.3, 1.3\}$ and magnitudes of $\beta_h = \{e^{-2.2}, e^{-2.5}, e^{-2.3}\}$ for the spatial periods $\lambda = \{1.76, 1.96, 2.5\}$ mm, respectively.

7.5.4 Influence of Grating Magnitude

The effect of the magnitudes of the gratings on the spatial spectra of the friction force fluctuations was studied with the three textures described earlier. Each grating was used to produce 100 measurements. The averaged results for all three gratings are plotted in Fig. 7.8.

The average spectra suggest that the magnitude of the surface undulation affects the amplitude of the fundamental component. The relationship is monotonic but probably not linear. Interestingly, however, the harmonic components do not seems to be affected by the gratings magnitude, at least within the range that we tested.

7.6 Discussion

The results presented above gives us a better understanding of the dynamic of the finger from a signal processing point of view and from the viewpoint of the mechanics involved.

Fig. 7.8 Spectrum averages
of 100 measurements made
with three gratings with same
spatial period of 1.96 mm and
variable amplitudes. Spectra
were normalized relatively to
their maxima in the low
frequencies, below the
frequency of the gratings

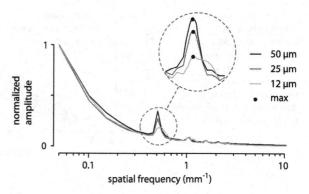

7.6.1 Background Noise

The $1/f$ background noise, present in the recordings, shed some light on the energy repartition of the vibration. This particular noise is found in many processes and has for particularity that the energy is constant in every decade band.

From a mechanical point of view, this trend can be explained by the fact that the height repartition of the smooth surface is close to white noise as its height density is a Gaussian function. By stroking our finger on the grating, this white noise drives a mechanical first order filter that represents a fingertip modeled by a spring and a dashpot.

It is worth noting that in signal generation, there is no simple way to spontaneously generate $1/f$ noise. The common approach to synthesize this signal is achieved by filtering random white noise with a first order filter.

7.6.2 Harmonic Behavior

The fact that the harmonic behavior of force fluctuations is different from the background noise cannot be explained simply. The decay of the harmonics is non-integer which does not fit with viscous nor with inertial dynamics. One possible element of answer is that the finger impedance and the friction conditions are strongly nonlinear, and therefore, the original sinusoidal forcing term becomes distorted and harmonics are created. The nonlinear transformation cannot simply be extrapolated from the measurements and requires analysis with more powerful tools such as Volterra series or Wiener kernels.

Another hypothesis can be formulated by supposing that the finger behaves like a distributed collection of oscillators forced by traction from the surface. Each of these oscillators undergoes stick-slip relaxation oscillations that are driven by the surface undulations. When one region of the skin sticks at one of the apices of the undulation, the skin is tangentially loaded, corresponding to a ramp in the interaction force. When the contact breaks away, energy is rapidly released corresponding

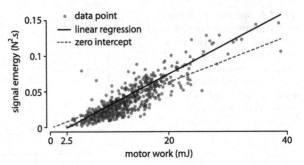

Fig. 7.9 Relationship between the work spent during a recording and the corresponding measured force signal energy. *Each dot represent one trial.* The first-order fit represents the data better than the zero-order fit ($R^2 = 0.88$ vs 0.76)

to a cliff. The force waveform associated with one element will be close to sawtooth waveform. The later picture is consistent with an interaction force signal exhibiting a decay of all the harmonic amplitudes since the harmonics of a sawtooth signal have amplitudes that are inversely proportional to their number. The occurrence of these oscillations would be hidden in the temporal domain by the superposition of several out-of-phase oscillators. This enticing possibility merits further investigations.

7.6.3 Energy Dissipation

Van den Doel et al. proposed a scaling law for the 'audio force' based on the instantaneous motor power delivered when scanning a surface (i.e. the square root of the product of tangential force and velocity) [22]. They assumed that the mechanical energy spent during the exploration is dissipated by friction and entirely transferred into propagating vibrations. This assumption is valid for rigid probe friction on rough surface, however a bare finger may behave differently. Viscoelasticity and soft tissue friction generate vibrations, but when stroking a finger on a smooth unlubricated surface, an elevation of skin temperature can be felt. It is the evident that some of the motoric work is transformed into heat.

For each trial, we estimated the work done by the participant to slide his finger and compared it to the vibratory signal energy. The motor work was calculated from $W = \int_0^d F_t(x)\,dx$, where d is the distance traveled, x is the net finger position and F_t is the tangential force component measured by the strain-gauge force sensor which is sensitive to DC and low frequencies components. The vibratory signal energy was determined from $E_s = \int_0^D |f_T(t)|^2\,dt$ where D is the duration of the movement, t is the time parameter and $f_T(t)$ is the force signal from the piezoelectric transducer since it is sensitive to frequencies above 2.5 Hz.

Figure 7.9 shows the relationship between the spent motor work and the signal energy for each trial. The linear regression with zero intercept of the vibration energy against motor work, shown in dashed grey, shows an acceptable fit with $R^2 = 0.76$. A first-order linear fit, shown in black, however, represents the data much better with $R^2 = 0.88$. The abscissa intercept is approximately at $W_{min} = 2.5$ mJ. This measurement, alone, does not inform us on the proportion of energy spent on heat and on vibrations. It nevertheless suggests that the work of bare-finger friction is transferred for the greatest part into vibrations originating at the finger-surface interface. The residual energy is dissipated into heat and probably radiated as sound waves.

The conditions that modify the surface-finger tribology are expected to greatly influence the proportions of the forms of dissipated energy, influencing in turn the information available for perceptual purposes. For instance, the vibrations arising from the frictional force can propagate in the surrounding air, giving acoustic cues, and in the skin giving tactile cues. Measurements of the finger and arm transmittance properties show that vibrations in the range 50 to 1000 Hz can travel as far as the forearm [8]. These results suggest that textures can possibly be sensed by very large populations of far flung mechanoreceptors located on the path of waves propagations.

7.7 Conclusion

We recorded accurate measurements of the force fluctuations generated by a finger exploring smooth flat and undulating surfaces. The spatial spectra of the signal shows that the vibrations are non deterministic but exhibit some interesting properties. With the smooth epoxy surface, we can characterize the background vibrations as a fractal noise with coefficient 0.9. The measurements with sinusoidal gratings reveal the nonlinearity of the skin interaction producing decaying harmonics compatible with the occurrence of a multiplicity of simultaneous relaxation oscillations.

These findings are the first steps toward the creation of virtual texture synthesis algorithms that are sensitive to various contact parameters and can generate appropriate vibrations. These findings can also possibly have perceptual implications since it can be proposed that the tactile system is sensitive to the peculiar dynamics involved. They also have possible contributions to make to the field of biotribology.

Future work will include finer experimentation on the influence of the material on the coefficient of the fractal noise, and its connection to the tribology and roughness. Moreover, other parameters such as the pose of the finger and hydration are likely to play dominant roles in the generation of tactual vibrations.

Acknowledgements The authors are grateful to A.M.L. Kappers for lending us the original gratings and Amir Berrezag for help on the casting procedure. They also would like to thank Yon Visell for his useful comments on the manuscript.

References

1. Adams MJ, Briscoe BJ, Johnson SA (2007) Friction and lubrication of human skin. Tribol Lett 26(3):239–253
2. André T, Lefevre P, Thonnard JL (2010) Fingertip moisture is optimally modulated during object manipulation. J Neurophysiol 103(1):402–408
3. Bergmann-Tiest WM, Kappers AML (2006) Analysis of haptic perception of materials by multidimensional scaling and physical measurements of roughness and compressibility. Acta Psychol 121(1):1–20
4. Billock VA, de Guzman GC, Kelso JS (2001) Fractal time and $1/f$ spectra in dynamic images and human vision. Physica D 148(1–2):136–146
5. Campion G, Hayward V (2008) On the synthesis of haptic textures. IEEE Trans Robot 24(3):527–536
6. Campion G, Hayward V (2009) Fast calibration of haptic texture synthesis algorithms. IEEE Trans Haptics 2(2):85–93
7. Costa MA, Cutkosky MR (2000) Roughness perception of haptically displayed fractal surfaces. In: Proceedings of the ASME dynamic systems and control division, vol 69, pp 1073–1079
8. Delhaye B, Hayward V, Lefevre P, Thonnard JL (2010) Textural vibrations in the forearm during tactile exploration. Poster 782.11. In: Annual meeting of the Society for Neuroscience
9. Guruswamy VL, Lang J, Lee WS (2010) IIR filter models of haptic vibration textures. IEEE Trans Instrum Meas 60(1):93–103
10. Hollins M, Faldowski R, Rao S, Young F (1993) Perceptual dimensions of tactile surface texture: a multidimensional scaling analysis. Percept Psychophys 54(6):697–705
11. Hollins M, Bensmaïa SJ, Washburn S (2001) Vibrotactile adaptation impairs discrimination of fine, but not coarse, textures. Somatosens Motor Res 18(4):253–262
12. Klatzky RL, Lederman SJ (1999) Tactile roughness perception with a rigid link interposed between skin and surface. Percept Psychophys 61(4):591–607
13. Krueger LE (1982) Tactual perception in historical perspective: David Katz's world of touch. In: Schiff W, Foulke E (eds) Tactual perception; a sourcebook. Cambridge University Press, Cambridge, pp 1–55
14. Mandelbrot BB (1982) The fractal geometry of nature. Freeman, New York
15. Nefs HT, Kappers AML, Koenderink JJ (2001) Amplitude and spatial-period discrimination in sinusoidal gratings by dynamic touch. Perception 30:1263–1274
16. Pataky TC, Latash ML, Zatsiorsky VM (2005) Viscoelastic response of the finger pad to incremental tangential displacements. J Biomech 38:1441–1449
17. Pawluk DTV, Howe RD (1999) Dynamic lumped element response of the human fingerpad. ASME J Biomech Eng 121:178–184
18. Persson BNJ (2001) Theory of rubber friction and contact mechanics. J Chem Phys 15(8):3840–3861
19. Siira J, Pai DK (1996) Haptic texturing-a stochastic approach. In: 1996 IEEE international conference on robotics and automation. Proceedings, vol 1, pp 557–562
20. Smith AM, Chapman CE, Deslandes M, Langlais JS, Thibodeau MP (2002) Role of friction and tangential force variation in the subjective scaling of tactile roughness. Exp Brain Res 144(2):211–223
21. Tomlinson SE, Lewis R, Carré JM (2007) Review of the frictional properties of finger-object contact when gripping. Proc Inst Mech Eng, Part B J Eng Tribol 221:841–850
22. Van Den Doel K, Kry PG, Pai DK (2001) FoleyAutomatic: physically-based sound effects for interactive simulation and animation. In: Proceedings of the 28th annual conference on computer graphics and interactive techniques. ACM, New York, pp 537–544
23. Wang Q, Hayward V (2007) In vivo biomechanics of the fingerpad skin under local tangential traction. J Biomech 40(4):851–860
24. Warman PH, Ennos AR (2009) Fingerprints are unlikely to increase the friction of primate fingerpads. J Exp Biol 212:2016–2022

25. Wiertlewski M, Lozada J, Pissaloux E, Hayward V (2010) Causality inversion in the reproduction of roughness. In: Kappers AML et al (eds) Proceedings of Europhaptics 2010. Lecture notes in computer science, vol 6192. Springer, Berlin, pp 17–24
26. Wiertlewski M, Hudin C, Hayward V (2011) On the $1/f$ noise and non-integer harmonic decay of the interaction of a finger sliding on flat and sinusoidal surfaces. In: World haptics conference (WHC). IEEE, New York, pp 25–30. doi:10.1109/WHC.2011.5945456
27. Wiertlewski M, Lozada J, Hayward V (2011) The spatial spectrum of tangential skin displacement can encode tactual texture. IEEE Trans Robot 27(3):461–472

Chapter 8
Conclusions

Abstract Recording and reproducing tactual textures with a high fidelity involves that the hardware and software are able to stimulate the skin at a sufficient speed and with a sufficient power. This chapter concludes the book by outlining the contribution made to the transducer design, to the understanding of the skin properties, and to the signal analysis procedure. It is followed by a discussion of the possible evolutions and improvements of this work.

8.1 Summary

The research reported in this book describes the first attempt, to my knowledge, to reproduce artificially the sensation of textured surfaces directly at the fingertip. It provide techniques for recording and reproducing finger-surface interaction forces accurately. These techniques required the investigation of the mechanical behavior of the fingertip in the range of displacement and frequencies relevant to touch as well as the influence this behavior on the reproduction of the vibrations caused by sliding on surfaces. The resulting biomechanical data are also useful for the design of transducers targeted at rendering textures at the fingertip and other applications. The mechanical signal measurements techniques and the analysis of the properties of these signals could be valuable tools for the design of realistic virtual textures rendering algorithms.

The main contributions address, (1) the design of new transducers for tactual interaction, (2) a study of mechanical phenomena subserving touch, and (3) the analysis of the vibration signal resulting from sliding a finger on a surface.

8.2 Transducers

Three electromechanical transducers were developed. The first was intended to provide for the accurate measurement of the finger-surface interaction forces, the second for the reproduction of tactual textures, and the last for the measurement of the fingertip mechanical behavior.

M. Wiertlewski, *Reproduction of Tactual Textures*,
Springer Series on Touch and Haptic Systems, DOI 10.1007/978-1-4471-4841-8_8,
© Springer-Verlag London 2013

The interaction of a finger with a surface is a complex mechanical process that involves a variety of phenomena that depend on numerous parameters including the topography of the surface, the stress distribution at the interface, the moisture flow, and other factors. Psychophysics experiments, however, indicate the existence of a marked phenomenon of perceptual constancy. Thanks to this constancy, vastly different mechanical stimuli can result in similar percepts as far as fine textures are concerned.

Despite the complexity of the physical interactions taking place during sliding, spatially distributed forces at the interface seem to be integrated, and thus, only net variations of the force strongly can determine a percept of texture. As a result, the frictional force and its rapid variations generated by fingertip slip, can be a sufficient representation of the interaction.

A measurement bench was designed and implemented to capture this frictional force within a frequency band and with a resolution that are compatible with the discriminative abilities of the human perceptual system. Since mechanical events occur in time, and also in space, the net position of the finger was also acquired at a resolution that matches the lowest characteristic lengths accessible to human touch.

To reproduce recorded textures, a second apparatus was developed using the same piezoelectric transducer. This transducer was designed to be five orders of magnitude stiffer than the fingertip. As a consequence, the actuator imposed its displacement over the fingertip seen as a load. By this technique, interaction with a surface was reproduced by shearing the pulp of the fingertip by a known quantity. Modulating the amount of shearing as a function of net finger position relatively to pre-recording data elicited a salient sensation of roughness. Several psychophysical experiments assessed the realism of this type of reproduction using roughness judgements, texture recognition and spatial period estimation.

This devices developed were designed for experimental purposes and were rather cumbersome. A miniaturized version of the tangentially skin shearing device is presently under development. This device can achieve up to 2 mm of deflection and is capable of stimulation in the plane rather than along one single direction,

Another apparatus was designed to quantify the mechanical impedance of the fingertip. The knowledge of this function allows one to predict the force to produce a specific velocity for any frequency. It is hope that the knowledge of this property will help clarifying tactile mechanisms taking place during sliding. It also provides us with a model of the fingertip that can be used to convert force measurements to displacement for purpose of tactual reproduction of virtual interactions.

This apparatus employs a novel approach to mechanical impedance measurement testing that draws its inspiration from the resonant measurements systems used in nanotechnologies and in other areas. It is built around a voice-coil actuator and uses wide-bandwidth analog feedback to reduce its apparent impedance by closed loop control. It is possible to lower the probe impedance sufficiently to match the impedance of the finger, thereby maximizing the exchange of mechanical energy, and hence, the signal-to-noise ratio of the measurement.

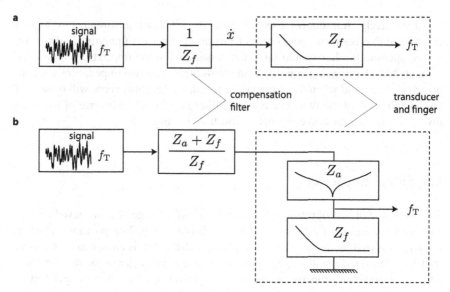

Fig. 8.1 Actuator impedance compensation. **a** In the case of a high impedance actuator. **b** In the case of a actuator which impedance is closed fingertip

8.3 Skin Mechanics

It was found that the bulk impedance of the fingertip in lateral traction exhibits an elastic behavior in the low frequencies and after a corner frequency of about 100 Hz, exhibit essentially a viscous behavior that prevails both the elasticity and inertia. This bimodal behavior might explain the decrease of the spatial frequency resolution of human cutaneous perception when temporal frequency increases. It is likely that this bulk behavior is the result of the structure of the fingertip that is nothing else but a fluid-saturated porous medium. In such media, the propagation of mechanical waves are disturbed and are frequency-dependent. High spatial frequencies and interferences are not preserved in the high temporal frequencies. When stimulated in by a complex surface, the skin would still undergoes bulk displacement that preserves the temporal information of the stimulation.

The knowledge of the fingertip impedance is equally important for transducer design. In audio technologies, all transducers work against similar loads—i.e. the impedance of free air if the wall are far enough. Measuring pressure, velocity or acceleration is a matter of proportionality. In the case of vibrotactile transducers, the load impedance is the skin being excited, and the transduction depends on frequency. Over the frequency band relevant to tactile perception, the skin behave as a first order mechanical filter. As a result, if a high-impedance transducer is used it is essential to convert a measured force signal into a displacement signal which implies the use of an inverse filter. Such filters should be included in the tactile signal processing pipelines in order to ensure the integrity of the vibrotactile signal perceived by the user, see Fig. 8.1a.

Alternatively, transducers used to generate vibrotactile stimuli can have a impedance that is of the same order of magnitude than the fingertip. In this case, the compensation filter should take into consideration the impedance bridge that is created by the fingertip coupled to the transducer. The two impedances are connected in series and when the actuator is excited, the resulting force will depend of the combination of the two. The resulting filter should be the inverse of the force divider created by the two elements, as illustrated in Fig. 8.1b.

8.4 Signal Encoding

The modeling of the vibrotactile behavior of a sliding finger was achieved by representing the signal as a function of finger displacement. Interpolation in spatial domain localized mechanical events, a property that touch is known to be sensitive to. Using this representation, slip velocity became a parameter of the measurement and not an intrinsic property of the signal. In fact, even though the finger-texture interaction measurements were made at an almost constant velocity, variations persisted which resulted in a shift of temporal frequencies of the signal that blurred the temporal spectrum.

This signal transformation privileged the spatial features of a surface and the short-term Fourier transform became a useful tool to analyze the properties of the sliding friction. The representations of complex textures in the spatial domain revealed an organization of the signal made around what could be called "spatial formants". In audition, temporal formants in a sound signal are known to contribute to the classification of sounds and speech. In touch, the discrimination of the temporal spectral organization is known to be quite poor, whereas the opposite is true for spatial organization. It is therefore possible to surmise that spatial spectral organization is an important perceptual cue for texture recognition even if the signal is sensed temporally.

It is well known that the perception of texture requires motion of the fingertip relatively to the surface to generate temporal vibrations. I was found however, that when a fingertip slides on simple undulated surface, the spatial frequency spectrum revealed a complicated non-linear transformation from the profile into tactile vibrations. We can therefore conclude that it no trivial matter to transform a surface into a tactile signal for synthesis and, conversely, to extract surface characteristics from a vibration for perception of its properties.

8.5 Future Work

The main result of this work is a complete tactile analysis/reproduction system that shears the fingertip as the finger undergoes a net relative motion. The method has shown its efficacy to reproduce fine virtual textured surfaces and its simplicity makes it possible to integrate miniature devices in more complex virtual environment systems in order to enhance the immersion.

The high impedance approach used in this work for the reproduction of textures is effective but require a powerful and stiff actuator. It would be possible to use a low impedance actuator in conjunction with a closed loop system. The control could then be tuned to emulate a stiffer system by mean of a fast position feedback. Such system would have to advantage to enabling the control adjust the transducer's impedance to match that of the fingertip. Doing so would optimize the power transfer from the actuator to the skin. A further advantage would be to make it possible to include in the feedback system the mechanical impedance bridge compensation in order to ensure of the integrity of the signal transmitted to the finger. Lower impedance of the actuators would also have the likely benefit of providing more amplitude from skin displacement.

Participants reported that the sensations elicited by the apparatus included roughness, but not the stickiness of the original textures. In fact, low frequencies and DC components of the frictional force were removed during reproduction to avoid actuator saturation. This reduction is probably the cause of the absence of stickiness sensation since the fingertip is displaced around a zero-mean value during reproduction of virtual textures. However, even if the net force taken into account, there is still the question of its perception. A more complete virtual reality system would unavoidably need to reproduce stickiness sensations as well.

To investigate the reproduction of stickiness, the vibrotactile transducer could be coupled to a force-feedback. Similarly to a tweeter and a boomer in a loudspeaker system, the low frequencies and high forces would be provided by force-feedback and high frequencies by vibrotactile transducers embedded in the manipulandum. The large inertia and artifacts arising from the force feedback devices could then be overcome by combining the two approaches.

Stickiness is not the only sensation that cannot robustly be rendered by vibrations. Sliding over virtual surfaces was sometimes experienced during experiments but often not. Some participants were able to feel that their finger was fixed to the transducer. The mechanical stimulus responsible for given the sensation sliding is not clearly identified, but it seems that vibrations can also mediate it. In fact, a simple experiment can be invoked to support this hypothesis. The sliding motion an object through a light mechanical filter such as a sheet of paper is still perceptible although the spatial information is strongly blurred. To our knowledge, a complete psychophysics and mechanical investigation of the sliding sensation have not been done yet but is necessary to lead to haptic interfaces able to provide more realistic simulations.

Printed in the United States
By Bookmasters